Formula One MATHS

€URO EDITION

PRACTICE BOOK

Yvonne Gostling ● **Mike Handbury**

Bob Hartman ● **Howard Johnson**

Jean Matthews ● **Colin White**

SERIES EDITOR: Roger Porkess

HODDER
EDUCATION
AN HACHETTE UK COMPANY

Acknowledgements

The authors and publishers would like to thank the following companies, agencies and individuals for permission to reproduce copyright material:

Map credits

p.5 Reproduced by permission of Ordnance Survey on behalf of HMSO. © Crown copyright 2006. All rights reserved. Ordnance Survey Licence number 36470

Every effort has been made to trace all copyright holders, but if any have inadvertently been overlooked the publishers will be pleased to make the necessary arrangements at the first opportunity.

Although every effort has been made to ensure that website addresses are correct at time of going to press, Hodder Education cannot be held responsible for the content of any website mentioned in this book. It is sometimes possible to find a relocated web page by typing in the address of the home page for a website in the URL window of your browser.

Throughout this book, the official spelling of the words 'euro' and 'cent' have been adopted, as specified by the European Central Bank. This may be seen as departing from the usual English practice for currencies but is correct for the euro currency.

Orders: please contact Bookpoint Ltd, 130 Milton Park, Abingdon, Oxon OX14 4SB. Telephone: (44) 01235 827720. Fax (44) 01235 400454. Lines are open 9.00–5.00, Monday to Saturday, with a 24-hour message answering service. Visit our website at www.hoddereducation.co.uk.

© 2003, 2006 Yvonne Gostling, Mike Handbury, Bob Hartman, Howard Johnson, Jean Matthews, Roger Porkess and Colin White

First published 2003
by Hodder Education,
an Hachette UK company,
Carmelite House, 50 Victoria Embankment,
London EC4Y 0DZ

Euro edition published 2006

Impression number   10
Year   2018

Cover image Jacey, Debut Art
Cover design and page design by Julie Martin
Illustrations by Jeff Edwards and Maggie Brand
Typeset by Tech-Set Ltd, Gateshead, Tyne & Wear
Printed and bound by CPI Group (UK) Ltd, Croydon, CR0 4YY

A catalogue record for this title is available from The British Library

ISBN: 978 0 340 92866 0

Hachette UK's policy is to use papers that are natural, renewable and recyclable products and made from wood grown in sustainable forests. The logging and manufacturing processes are expected to conform to the environmental regulations of the country of origin.

Contents

1 How our numbers work

1.1 Large numbers

1 Write each of these numbers as a power of ten.
   (a) 1000   (b) 10 000   (c) 1 000 000   (d) 100 000 000 000

2 Write in words:
   (a) 100 000   (b) 1 000 000   (c) 10 000 000 000

3 How many noughts appear in the number ten million?

4 Write these numbers in figures.
   (a) Seven thousand and sixty three.
   (b) Twenty three thousand, six hundred and three.
   (c) Five hundred and seven thousand, two hundred and fourteen.
   (d) Eleven million, eleven thousand and eleven.
   (e) Ten billion, seventy two million, six hundred and seventeen thousand, two hundred and eight.

5 Write these numbers in order, starting with the smallest.
   10^9 one million   10 000 000 000   one thousand   100 000

6 The area of Lebanon is ten thousand square kilometres.
   Write ten thousand in figures.

7 The list below gives the numbers of cars stolen in five countries in one year.
   (a) Finland   21 043   (b) Italy   302 490   (c) Austria   4248
   (d) UK   592 685   (e) USA   1 539 287
   Write each of these numbers in words.

8 For each of the following numbers say which column the 4 is in, and how much it represents.
   (a) 4618   (b) 37 406   (c) 5 791 043   (d) 346 819 000

9 Which is longer, one billion seconds or 30 years?
   Show your working.

1.2 Using large numbers

1 Write the answers to the following:
   (a) 30×50 (b) 40×70 (c) 50×120 (d) 90×300
   (e) 810×30 (f) 920×400 (g) 300×6500 (h) $7500 \times 20\,000$

2 Write the answers to the following:
   (a) $700 \div 10$ (b) $240 \div 20$ (c) $900 \div 30$ (d) $3200 \div 40$
   (e) $1500 \div 50$ (f) $8000 \div 400$ (g) $36\,000 \div 900$ (h) $4\,980\,000 \div 6000$

3 Fill in the gaps in the following statements.
   (a) $50 \times \underline{\hspace{1cm}} = 3500$
   (b) $\underline{\hspace{1cm}} \times 80 = 7200$
   (c) $120 \times \underline{\hspace{1cm}} = 4800$
   (d) $\underline{\hspace{1cm}} \times 600 = 180\,000$
   (e) $\underline{\hspace{1cm}} \times 40\,000 = 14\,000\,000$
   (f) $36\,000 \div \underline{\hspace{1cm}} = 1800$
   (g) $400\,000 \div \underline{\hspace{1cm}} = 8000$
   (h) $\underline{\hspace{1cm}} \div 20 = 1600$
   (i) $\underline{\hspace{1cm}} \div 300 = 700$
   (j) $400\,000\,000 \div \underline{\hspace{1cm}} = 200\,000$

4 The distance from the Earth to the Moon is about 384 000 kilometres.
A spacecraft travels at an average speed of 5000 kilometres per hour.
Roughly how long does it take the spacecraft to travel to the Moon and back?

5 The distance around Saturn's equator is about 380 000 kilometres.
Saturn turns round once every $10\frac{1}{4}$ hours.
How fast does a point on the equator travel as Saturn spins around?
Write your answer in kilometres per hour and in metres per second.

6 In one orbit of the Sun, Mars travels about 712 million kilometres.
The orbit takes 687 days.
How fast does Mars travel as it orbits the Sun?
Give your answer in kilometres per hour.

7 The speed of light is 300 000 000 metres per second.
The distance of the Sun from the Earth is about 150 000 000 kilometres.
How long does it take for light to travel from the Sun to the Earth?

1.3   The metric system

1 Copy and complete these statements.

(a) 1 metre = _____ millimetres

(b) 1 _____ = 100 centimetres

(c) 1 _____ = 1000 metres

(d) 1 kilohertz = _____ hertz

(e) 1 litre = _____ centilitres

(f) 1 kilogram = _____ grams

(g) 1000 newtons = _____ kilonewton

(h) 1 kilogram = _____ milligrams

(i) 1000 millilitres = _____ litre

(j) 10 millilitres = 1 _____

2 Write each of these measurements in two different ways.

(a) 75 grams

(b) 45 centimetres

(c) 0.4 kilometres

(d) 25 centilitres

3 Write the following in order, smallest first.
$\frac{1}{2}$ litre   75 cl   450 ml   0.4 litres   2.5 cl

4 From the measurements below, select any distances that are equivalent.

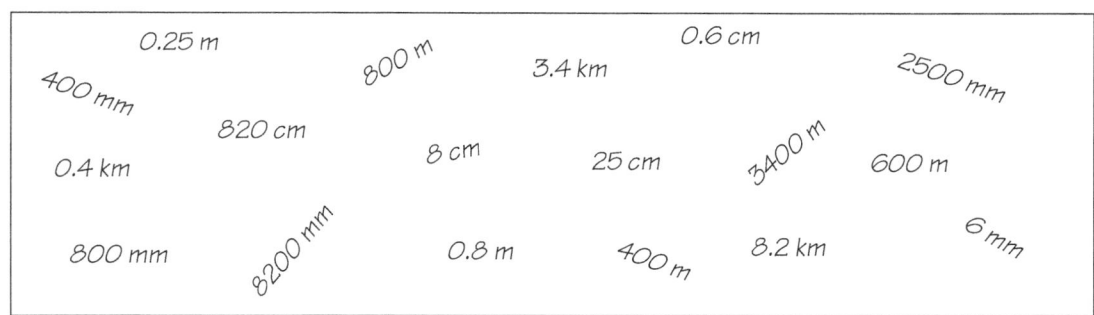

2 Position

2.1 Position

For this exercise you will need squared paper.

1 This is part of an electronic display board.
It is turned on to show the letter 'L'.

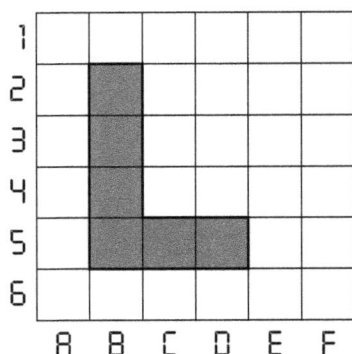

(a) Which three squares need to be turned on to change the 'L' into a 'U'?
(b) Which squares need to be turned on to change the 'U' into an 'O'?
(c) A full stop (.) is made from four squares.
The position of the top right-hand square is D4.
What are the positions of the other three squares?
(d) Which letter is made by these squares:

B5   B2   E5   E2   C4   C3   D4   D3?

2 For this question you will need to make copies of the grid in question 1.
Allow yourself about five minutes to write down the combinations of squares which will display digits.
See how many you can make in the time.

3 (a) Which of these display a horizontal bar?
   (i)   A6 B5 C4
   (ii) A2 B2 C2
   (iii) C4 C3 C2
   (iv) F1 F2 F3
   (v)   A2 B3 C4
   (vi) E4 D3 C2
   (b) Find the rule in the numbers and letters for horizontal bars.
   Test it for yourself.

2.2   The National Grid

1 Here is a sketch of part of the east of England coast showing National Grid squares.

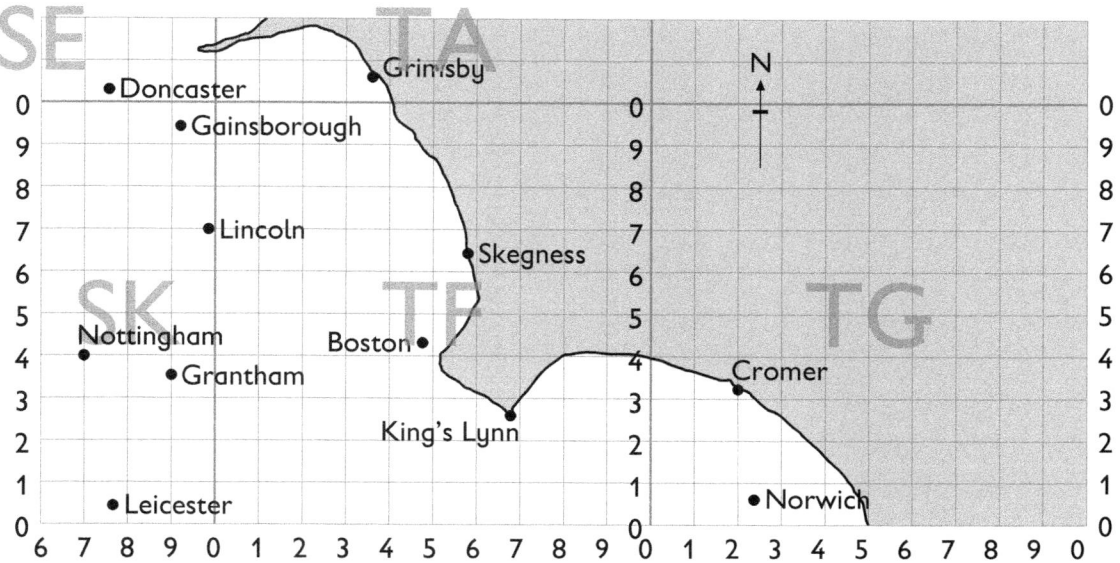

(a) What places are in these squares?
    (i)   SK70
    (ii) TF44
    (iii) TG23

(b) What is at TG67?

(c) Give the National Grid references of these places.
    (i)   Doncaster
    (ii) Norwich
    (iii) Grimsby

(d) The large lettered squares are 100 km by 100 km.
    (i)   What are the side lengths of the small squares?
    (ii) Which place on the map is about 60 km north of Grantham?
    (iii) Which place is about 150 km east of Leicester?

2 (a) Imagine two towns whose National Grid references are given by SY12 and SY14. What can you say about their positions?

(b) What can you say about the towns whose National Grid references are given by SY22 and SY99?

3 Check your answers to **1(a)** and **(c)** (and the sketch map!) by visiting the site
http://getamap.ordnancesurvey.co.uk/frames.htm

3 Basic number

3.1 Addition and subtraction

1 The table shows distances in kilometres between towns in Ireland.

Belfast						
403	Cork					
181	395	Donegal				
162	256	229	Dublin			
296	205	203	213	Galway		
325	101	294	195	104	Limerick	
318	125	346	158	218	126	Waterford

Find the distances travelled in the following journeys.
  (a) Belfast to Dublin then Dublin to Waterford
  (b) Donegal to Galway then Galway to Dublin
  (c) Cork to Limerick then Limerick to Dublin
  (d) Limerick to Galway then Galway to Belfast then Belfast to Dublin
  (e) Cork to Waterford then Waterford to Dublin then Dublin to Donegal.

2 James wants to save €475 to buy a guitar.
How much more does he need to save at each of these stages:
  (a) he has saved €142 at the end of the first month
  (b) he has saved €290 altogether at the end of the second month
  (c) he has saved €386 altogether at the end of the third month?

3 The top of a computer desk is supported by two
cupboards of different sizes.
The large cupboard has width 455 mm.
The small cupboard has width 210 mm.
Between the cupboards there is a 'kneehole' of
width 592 mm.
  (a) What is the total width of the desk?
  (b) The desk is to be placed in a recess of width
1500 mm. What is the width of the gap?

4 **Investigation**

In a game of darts players take turns to throw three darts at a circular dart board. The points scored by each dart thrown are subtracted from the starting score of 501. The first player to reach zero wins. Players can score single, double or treble 1 to 20; they can also score 25 for the outer bull and 50 for the bull's-eye.

In her first turn Amanda scores double 20, treble 12 and single 7. So her new score is 418.

Starting from her new score of 418 calculate Amanda's score at the end of each turn:
(a) in her second turn she scores double 18, single 12 and treble 9
(b) in her third turn she scores single 10, double 19 and bull's-eye
(c) in her fourth turn she scores treble 20, single 18 and single 13.

A player must end with an exact score of zero and the last dart must score a double or a bull's-eye.
(d) In how many different ways can Amanda finish the game in her fifth turn? Show these ways.

3.2 Short multiplication and division

1
Tickets for a coach trip to Eurocity via the Eurotunnel cost €43 each.
Find the total cost of:
(a) four tickets bought by a family
(b) 40 tickets bought by the passengers on one coach
(c) 200 tickets bought by the passengers on five coaches.

2
What is the total cost of show tickets for:
(a) a family of four in Section D
(b) five friends in Section C
(c) a mini-bus party of 15 in Section B
(d) a coach party of 28 in Section A?

End of the Pier Show.	
Ticket prices	
Section A	€5
Section B	€6
Section C	€8
Section D	€12

3   To convert kilometres to miles, divide by 8 and multiply by 5.
Convert the following into miles.
(a)   16 km   (b)   40 km   (c)   24 km   (d)   80 km   (e)   64 km

4   Suzanne is organising a disco.
How much should she charge for each ticket so that she can cover her costs after selling 40 tickets?

Hire of the hall	€50
Disco	€150
Refreshments	€72
Printing	€8

5   ## Activity

Copy and complete this crossnumber.

Across	Down
1 3 × 128	**1** 4 × 784
4 78 ÷ 3	**2** 5 × 172
6 8 × 212	**3** 7 × 701
7 25 × 12	**5** 2496 ÷ 4
8 336 ÷ 4	**8** 3 × 2839
9 5 × 147	**10** 156 ÷ 4
12 2 × 456	**11** 470 ÷ 5
14 4 × 235	**13** 104 ÷ 4
15 380 ÷ 5	

3.3   Long multiplication and more division

Show your working in every question.

1   A hotel charges €55 per night.
How much would it cost to stay:
(a)   6 nights   (b)   10 nights   (c)   16 nights?

2   Boarding kennels charge €14 per dog per day.
Calculate the charge for:
(a)   1 dog staying for 14 days   (b)   2 dogs staying for 12 days
(c)   2 dogs staying for 16 days   (d)   3 dogs staying for 15 days.

3   Here are the amounts that some people are paid.
What is the weekly wage in each case?
(a)   €1008 for 8 weeks   (b)   €1344 for 12 weeks
(c)   €1890 for 14 weeks   (d)   €4032 for 28 weeks

4   What is the daily cost of:
(a)   an 8-day holiday costing €672   (b)   a 12-day holiday costing €1020
(c)   a 14-day holiday costing €1344   (d)   an 18-day holiday costing €1584?

5 **Investigation** ──────────────────────────────

Last digits
$8 \times 4 = 32$, so the *last digit* of 8×4 is 2.
(a) What is the last digit of:
    (i) 18×4 **(ii)** 28×14 **(iii)** 68×24 **(iv)** 98×94?
    What do you notice about your answers?
(b) What is the last digit of:
    (i) 7×9 **(ii)** 17×19 **(iii)** 67×39 **(iv)** 97×49?
(c) Look at all your answers to **(b)**. Now state a rule for finding the last digit without
    using long multiplication.
    Use your rule to find the last digit of:
    (i) 89×19 **(ii)** 119×59 **(iii)** 139×49
    (iv) 39^2 **(v)** $19 \times 29 \times 39$ **(vi)** 59^3
(d) Without using a calculator or doing long multiplication state which of these
    answers must be wrong. Give a reason for your answer.
    (i) $37 \times 46 = 1512$ **(ii)** $68 \times 42 = 2854$ **(iii)** $54 \times 23 = 1244$
    (iv) $96 \times 32 = 3072$ **(v)** $12^3 = 1724$

3.4   Multiples and divisibility tests

1 Which of the following numbers are exactly divisible by 5?
    305   630   845   432   777   574   935   1055

2 Which of the following numbers are exactly divisible by 3?
    51   43   87   79   114   7032   5133   6240

3 Which of the following numbers are exactly divisible by 4?
    328   424   202   526   8024   312   3164   24 672

4 If a number can be divided exactly by 3 *and* by 4, then it can be divided by 12, because
    $3 \times 4 = 12$.
    Which of these numbers are multiples of 12?
    408   452   624   568   742   864   1128   1344

5 Bonzo dog food is sold in two sizes – standard
and large.
Standard cans are packed in boxes with 18
cans to each layer.
Large cans are packed in boxes with 12 cans
to each layer.
Look at the number of cans in these cartons
and state whether each carton contains
standard or large, or whether you cannot decide.
Show the divisibility test you used to decide
your answer.

6 What is the remainder when 32 419 is divided by:

(a) 3   (b) 4   (c) 5   (d) 9?

7 Here are three numbers.

6329   8542   9563

Fit the numbers to these descriptions.

(i) When divided by 3 the remainder is 2. When divided by 4 the remainder is 3. When divided by 5 the remainder is 3.	**(ii)** When divided by 4 the remainder is 2. When divided by 5 the remainder is 2. When divided by 9 the remainder is 1.

(iii) When divided by 3 the remainder is 2.   When divided by 4 the remainder is 1.   When divided by 5 the remainder is 4.

8 What is the smallest number which satisfies these statements?

When divided by 3 the remainder is 1.

When divided by 4 the remainder is 2.

When divided by 5 the remainder is 4.

3.5   Why the divisibility tests work

1 Numbers that are divisible by 3 *and* by 5 are divisible by 15.

State why you cannot find numbers that are divisible by 8 by checking that they are divisible by 2 *and* by 4.

2 Devise a divisibility test for

(a) 18

(b) 45

3 A number can be divided by 4 exactly if 4 goes exactly into the last two digits.

A number can be divided by 8 exactly if 8 goes exactly into the last three digits.

State a test for divisibility by 16.

4 Which of these numbers is a multiple of (a) 3   (b) 11   (c) 33?

     28 875   47 346   49 896

Devise a divisibility test for 33.

5 Mr Smythe the builder has given James a bill for some work he has done.

Unfortunately James dropped the bill in a puddle and he can't read all the figures.

James knows the hourly rate is less than €20.

Use divisibility tests to work out the missing figures.

Labour.
45 hours @ €?? = €?9?

6 **Investigation**

In this number A and B represent missing digits: 5A1B2.

Find all the possible values of A and B so that this number is a multiple of 11.

4 Angles

4.1 Types of angle

You will need a protractor for this exercise.

1 Copy these columns and join each angle to its correct description with a line.

30°	reflex
300°	acute
90°	acute
120°	reflex
60°	right-angle
205°	obtuse

2 **(a)** Write these angles in order of size, smallest first.

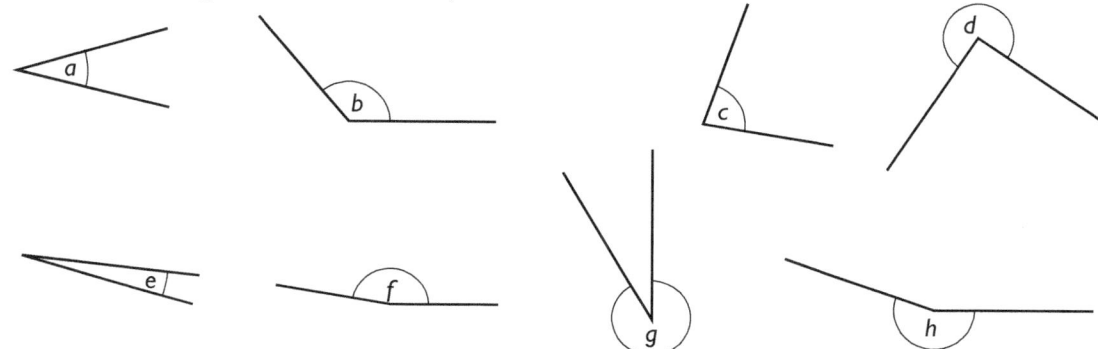

(b) Measure the angles to see if you were correct.

(c) Copy and complete this table by putting each angle letter in the correct column.

Acute	Obtuse	Reflex

3 **(i)** 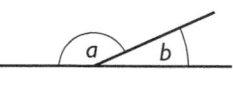   **(ii)** 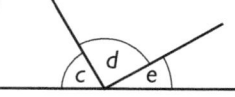   **(iii)**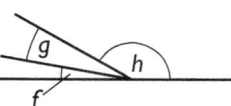

In each part:

(a) Measure the angles. Write the size of each angle. Then write whether it is acute, right-angled, obtuse or reflex.

(b) What do you notice if you add the angles together?

4 **Investigation**

(a) Measure the angles *a*, *b*, *c* and *d*.

(b) Add these together. What do you notice?

(c) Try the same for other quadrilaterals. What do you notice?

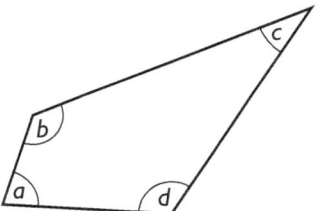

4.2　Drawing angles

You will need a protractor and a ruler for this exercise.

1 Draw these angles accurately using a protractor.

(a) 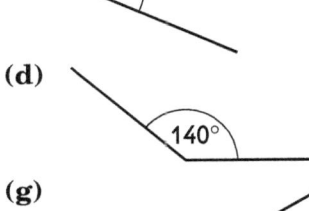 **(b)** 65° **(c)** 73°

(d) 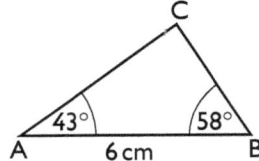 **(e)** 115° **(f)** 167°

(g) 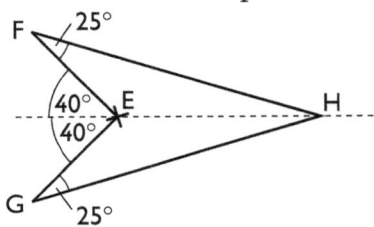 **(h)** 256° **(i)** 310°

2 Construct this triangle accurately and measure the angle at C.

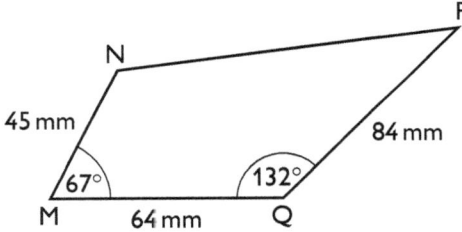

3 Construct this quadrilateral accurately.

Measure the angles at N and P and the side NP.

4 Construct this shape accurately.

EF = EG = 4 cm

5 **(a)** Draw a circle of radius 3 cm and mark the centre.
Draw six radii at an angle of 60° to each other.
Join the ends of these radii. You have constructed a hexagon.
(b) Repeat **(a)** with an angle of 45° between the radii.
What shape have you constructed?
Try other angles.

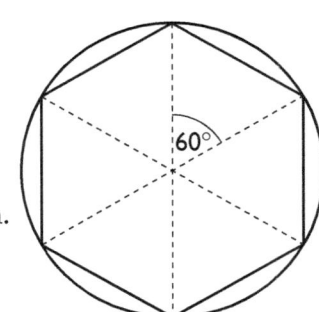

4.3   Angle calculations

1   Work out the value of each lettered angle in these diagrams. Write the angle fact that you use in each case.

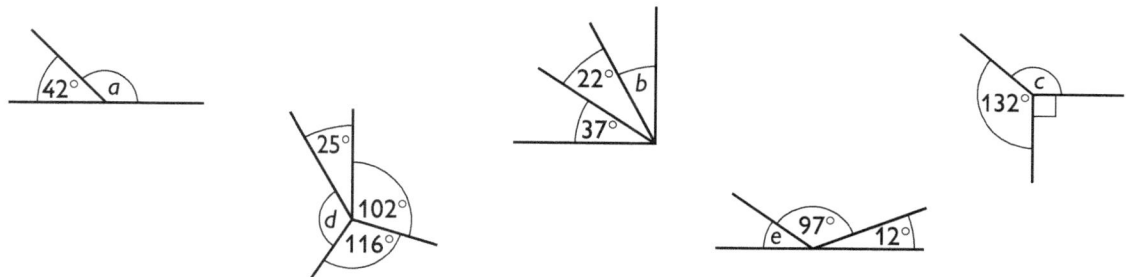

2   Make a sketch of each of these diagrams. Find the value of each lettered angle and give reasons for your answers.

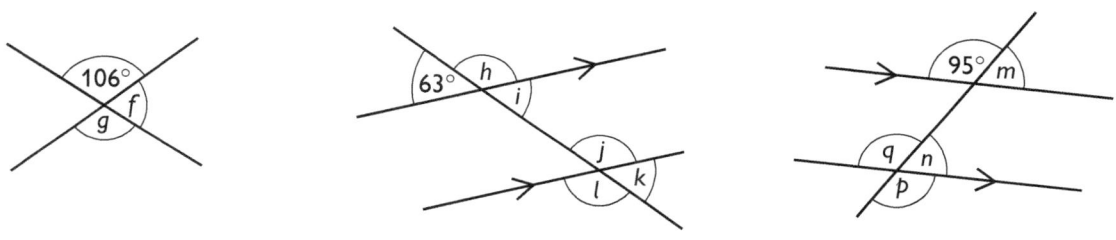

3   Make a sketch of each of these diagrams.
Find the value of each lettered angle and give
a reason for your answer.

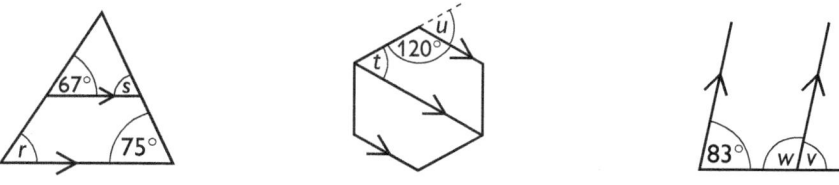

4   Make a sketch of each of these diagrams.
Find the value of each lettered angle and give
a reason for your answer.

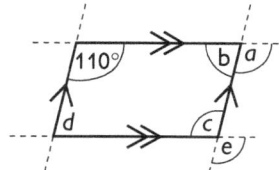
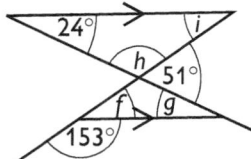

4.4   Angles in a triangle

1   Find the value of each lettered angle in these diagrams.

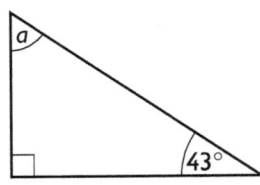
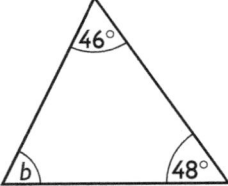
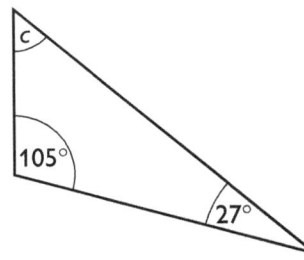

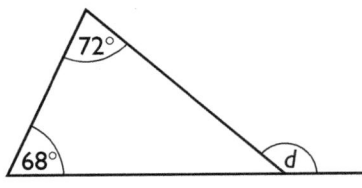
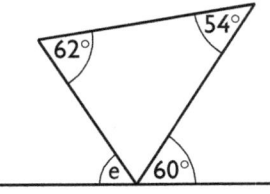
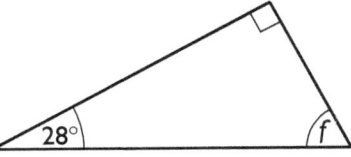

2   Find the value of each lettered angle in these diagrams. Give a reason for your answer.

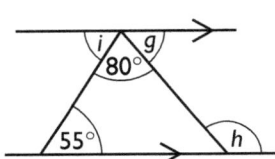
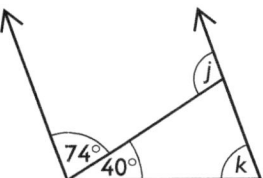

3   Find the value of each lettered angle in these diagrams. Give a reason for your answer.

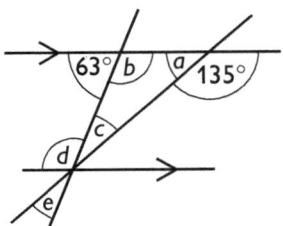
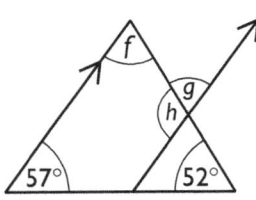

5 Displaying data

5.1 Displaying data

1 Maria threw a die 40 times.
She kept a tally chart of her results.
   (a) Complete the frequency column in her table.
   (b) Which number came up most often?

Score	Tally	Frequency
1	⊬ I	6
2	⊬ III	
3	⊬ ⊬	
4	⊬ II	
5	⊬ I	
6	III	

2 Sarah is doing a survey about school lunches. She asks 30 people in the lunch queue what their favourite lunch is. These are her results.

pizza	chips	pizza	salad	sandwiches
sandwiches	hotdogs	burgers	chips	salad
pizza	hotdogs	salad	sandwiches	pizza
salad	hotdogs	burgers	pizza	burgers
chips	sandwiches	hotdogs	hotdogs	pizza
pizza	salad	burgers	sandwiches	sandwiches

   (a) Make a tally chart of Sarah's results.
   (b) What is the most popular food?
   (c) What is the least popular food?

3 Investigation

Anna did an experiment with an ordinary pack of playing cards. She drew a card, noted whether it was an ace, a picture or a number using A, B or C. She repeated this 40 times. Here are her results.

A	A	B	C	B	B	B	C	A	A
A	A	C	C	A	B	A	C	C	C
A	C	A	A	A	A	C	B	A	B
A	A	B	A	C	C	A	A	A	A

Make a tally chart of these results to show how many of each card she drew. Use your tally chart to decide which letter represents:
   (a) an ace   **(b)** a picture card   **(c)** a number card.

5.2 Pictograms and bar charts

1 John asked all the students in his class 'How many children are there in your family?'
This pictogram shows the results.

1 child	○ ○ ○ ◖
2 children	○ ○ ○ ○ ○
3 children	○ ○ ◖
4 children	○
5 children	○
6 children	◖

Key
○ = 2 families

(a) How many families had two children?
(b) How many families had three children?
(c) How many students are in John's class?
(d) Draw a bar chart to show these data.

2 Hannah asked her friends about their favourite TV 'soap'.
She drew a bar chart to show the results.
(a) What is the most popular programme?
(b) How many of her friends prefer Emmerdale?
(c) How many of her friends prefer EastEnders or Hollyoaks?
(d) How many friends did she ask all together?
(e) Draw a pictogram to show these data.

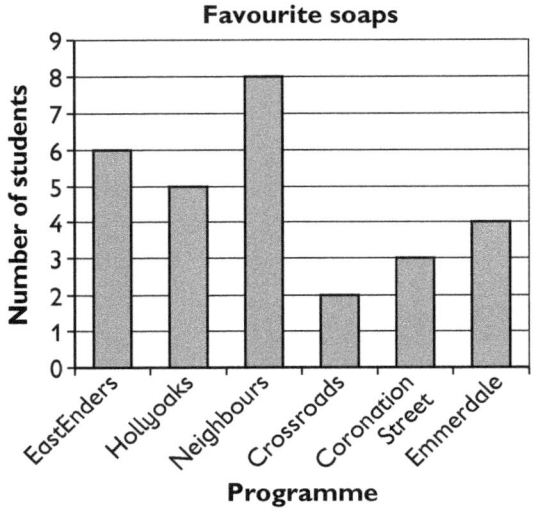

3 Ashley is doing a survey to find out the most popular sport on TV.
This frequency table shows the results.

Sport	Football	Tennis	Golf	Snooker	Rugby
Frequency	12	6	4	4	2

(a) Draw a pictogram to illustrate these data.
(b) Draw a bar chart to illustrate these data.

4 **Investigation**

Ellie did a survey of eye colour.
She drew this pictogram to show her results.

Blue eyes   ○ ○ ○ ○

Brown eyes   ○ ○ ○ ⊂

Grey eyes   ○ ○ ◖

Green eyes   ○ ⊂

Ellie forgot to add a key to her pictogram.
(a) What is the smallest number of people who could have taken part in the survey?
(b) Ellie spoke to less than 150 people. What is the biggest number of people who could have taken part in the survey?

5.3   Pie charts

1 The pie chart shows the colour of sweets in a packet of fruit drops.
State whether each of these statements is true or false.
(a) There are more orange sweets than purple sweets.
(b) There are fewer green sweets than yellow sweets.
(c) A quarter of the sweets are red.
(d) There are more purple sweets than red sweets.
(e) Half of the sweets are orange or purple.

Colour of sweets

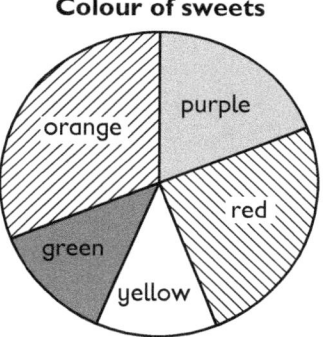

2 Dominic did a survey on where the pupils in his class spent their last summer holiday.
One half of the class had its holiday in Britain.
One third of the class had its holiday in Spain.
The rest had their holiday in France.
Draw a pie chart to show this information.

3 What fractions correspond to a pie chart angle of:
(a) 45°   **(b)** 135°
(c) 270°   **(d)** 240°?

4 What pie chart angle corresponds to a fraction of:
(a) $\frac{1}{6}$ **(b)** $\frac{2}{3}$
(c) $\frac{1}{9}$ **(d)** $\frac{5}{8}$?

5   Lauren has conducted a survey with 24 boys and 24 girls to find out their favourite fruit. Here are the pie charts she has drawn.

Boys

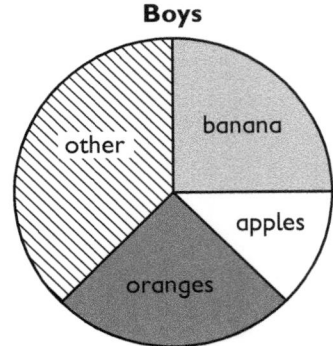

Girls

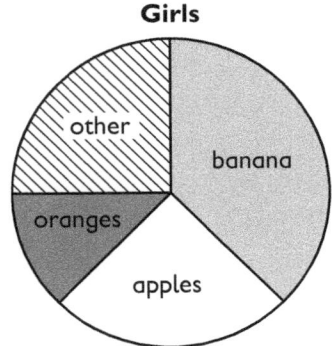

Find the number of boys who prefer each fruit. Write them also as fractions. Now give the same information for the girls.

5.4   Grouping data

1   30 people did an experiment in which each person had to estimate when a minute had passed while their partner timed them. These are their results in seconds.

50	51	50	52	55
59	75	62	59	56
46	65	70	52	60
61	62	52	66	62
72	63	57	47	60
62	57	58	49	66

Time	Tally	Frequency
46–50		
51–55		
56–60		
61–65		
66–70		
71–75		

(a)   Copy and complete this tally chart.
(b)   Draw a bar chart of the results.

2   In the 1881 Census these were the ages of the residents of a small village.

36	33	11	9	6	3	60	56	9	17
17	16	49	59	3	24	23	67	66	69
55	51	17	12	10	6	25	40	36	15
13	10	8	6	4	2	1	75	46	42
11	8	5	4	56	49	44	7	37	4

(a)   Make a tally chart using groups 0–9, 10–19, and so on.
(b)   Draw a bar chart.
(c)   How many people lived in the village?
(d)   Which age group had the highest number of residents?

6 Symmetry

6.1 Symmetry

1 Copy these shapes and draw all their lines of symmetry. Write down how many lines of symmetry each one has.

(a) 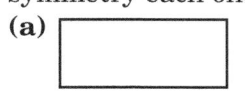 (b) 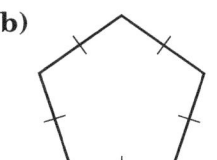 (c) 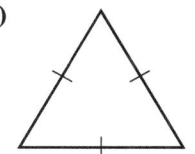 (d) 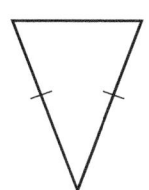 (e)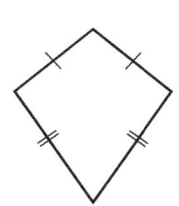

2 (a) Copy these car badges and draw all their lines of symmetry. Write down how many lines of symmetry each one has.

(b) Design a car badge of your own and draw all the lines of symmetry.

3 Copy each shape and the line of symmetry. Draw the rest of each symmetrical shape.

(a) 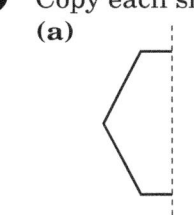 (b) 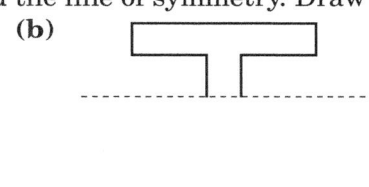 (c)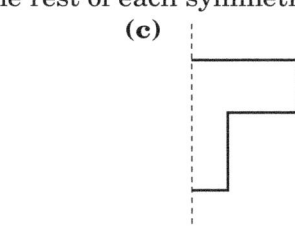

4 Copy each shape and the line(s) of symmetry. Draw the rest of each symmetrical shape.

(a) 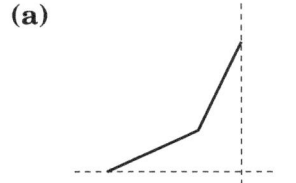 (b) 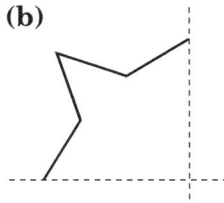 (c)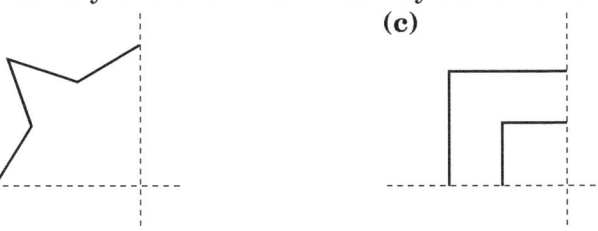

(d) 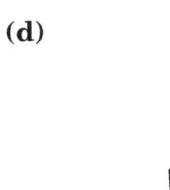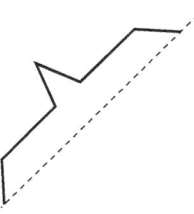 (e)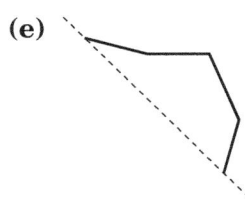

5 The diagrams below show patterns and their line(s) of symmetry. The patterns are incomplete. Copy the diagrams, then draw in the missing black squares.

(a)
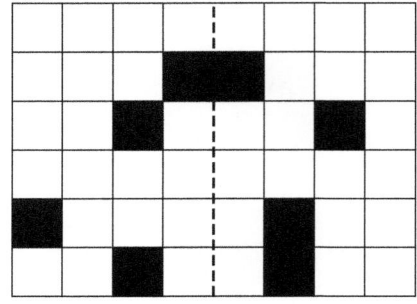

(b)
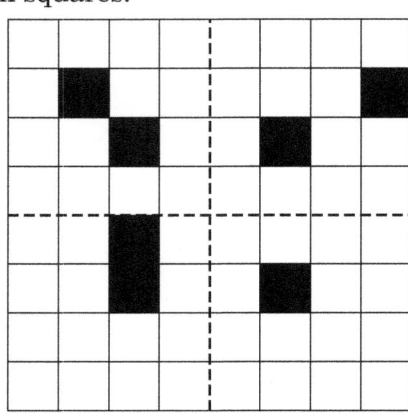

6.2   **Rotational symmetry**

1 Copy each of these shapes and mark its centre of rotation with a cross. Write down its order of rotational symmetry.

(a)
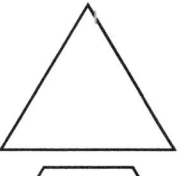

(b)

(c)
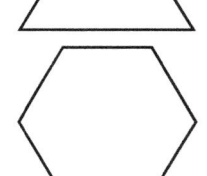

(d)
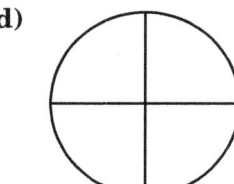

(e)
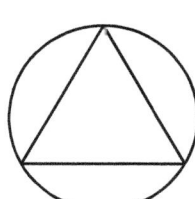

(f)
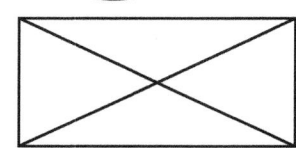

2 Copy each of these shapes. The centre of rotation is shown with a cross. Draw the rest of the shape. The order of rotation is given.

(a)
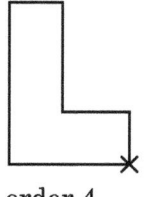

order 4

(b)
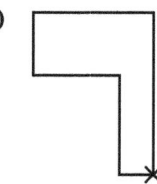

order 2

3 Trace each of these shapes. For each one:
   (i) Write down how many lines of symmetry it has and draw them.
   (ii) Write down the order of rotational symmetry and mark the centre with a cross.

   (a) 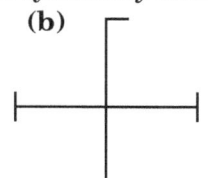   **(b)**   **(c)**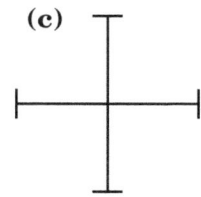

4 Trace each of these shapes. For each one:
   (i) Write down how many lines of symmetry it has and draw them.
   (ii) Write down the order of rotational symmetry and mark the centre with a cross.

   (a) 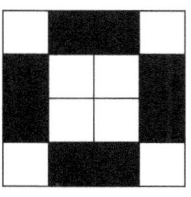   **(b)** 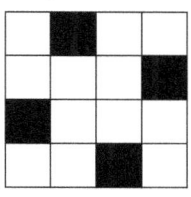   **(c)**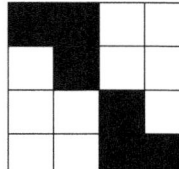

   (d) 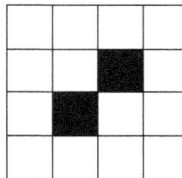   **(e)**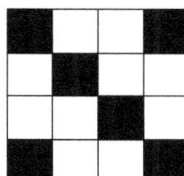

6.3 Paper folding

For this exercise you will need some pieces of A4 paper to fold.

1 **(a)** Fold a piece of A4 paper twice and cut an isosceles triangle from the corner. When you open the paper, what shape have you cut out?

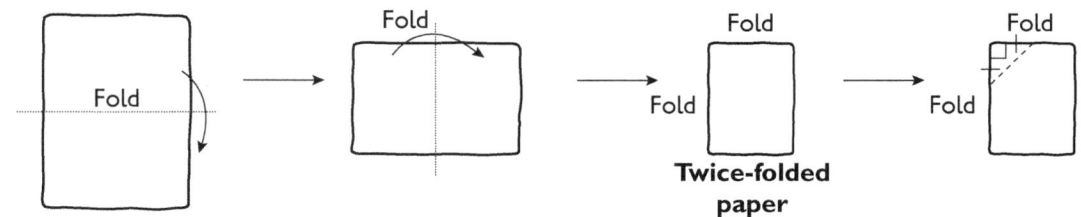

Twice-folded
paper

   (b) Draw four copies of the twice-folded paper and show the cut needed to get these shapes:
   (i)   rhombus
   (ii) octagon
   (iii) rectangle
   (iv) circle.

2 A piece of A4 paper is folded twice and an isosceles triangle is cut from each corner. When you open the paper what will it look like?

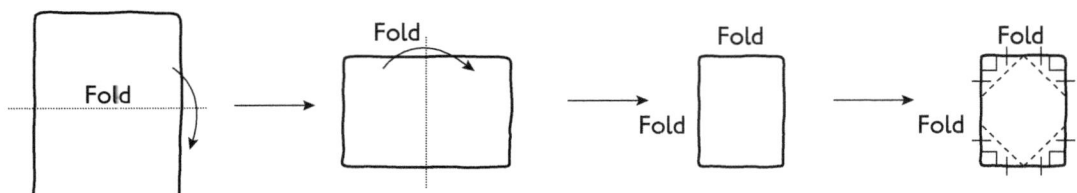

3 Henry folds a piece of A4 paper twice and cuts some shapes out of it.
When it is open it looks like the diagram below.
Show the cuts that Henry made on the twice-folded paper.

 **Decimals**

7.1   Decimals

1 Copy and complete these.
The first has been done for you.
  (a) 4.2 is 4 units + 2 tenths.
  (b) 14.6 is _____
  (c) 10.5 is _____
  (d) 101.4 is _____
  (e) 17.0 is _____

2 Write these numbers in order of size:

  0.9   3.3   0.3   0.7   3   1.1

3 Find the volume of water in each of these jugs.
  (a)   **(b)**   **(c)**

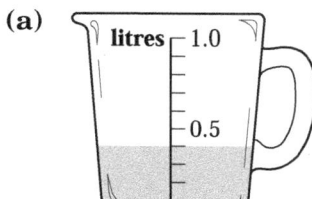      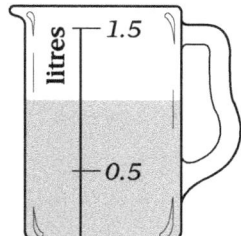

4 What numbers are these arrows pointing to?

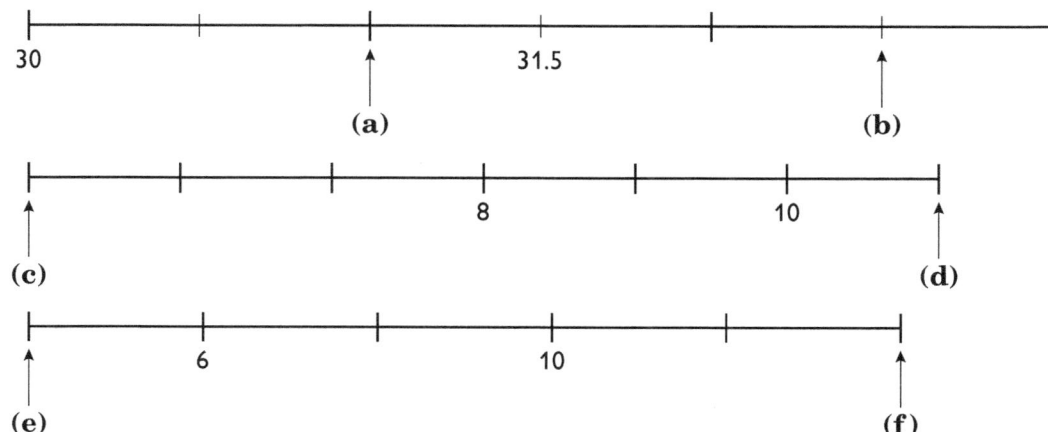

5 A single graduation is called a tick, for example this scale has
two ticks between 1 and 2.
A scale is graduated in divisions of 0.1.
How many ticks are there in-between 2.0 and 3.0?

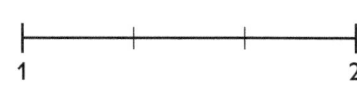

7.2   Decimal notation

1 Write down the value of the 4 in each of these numbers:
  (a) 140   **(b)** 3.45   **(c)** 5.34
  (d) 103.4   **(e)** 0.045.

2 Write the next two numbers for each of these number patterns:
  (a) 1.2   1.4   1.6   1.8
  (b) 1.92   1.94   1.96   1.98
  (c) 4.68   4.71   4.74   4.77.

3 Amber reads 0.16 as 'zero point sixteen'.
  Why is Amber wrong?

4 What numbers are these arrows pointing to?

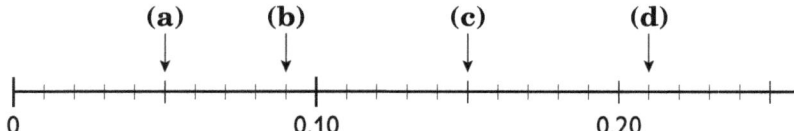

5 Write these numbers as decimals:
  (a) a half   **(b)** two and a half   **(c)** three tenths
  (d) seven hundredths   **(e)** seventy thousandths   **(f)** seven thousandths.

6 Sketch a number line from 0 to 2, like this one.

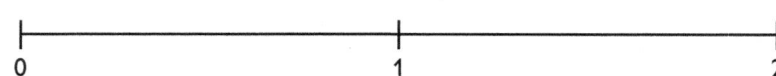

  Estimate the position and mark these numbers with letters:
  (a) 0.5   **(b)** 1.50   **(c)** 1.05.

7 Write one or two sentences to convince a friend that 1.39 is smaller than 1.5.
  You may find it helpful to use diagrams.

8 **(a)** What decimal is one-tenth of the way from 0 to 1 on the number line?
  (b) What decimal is one-tenth of the way from 0.1 to 0.2 on the number line?
  (c) What decimal is one-tenth to the left of 1 on the number line?
  (d) What decimal is one-hundredth to the left of 1 on the number line?

9 Using a calculator to help you, put these fractions in order, smallest first:
  $\frac{11}{15}$ $\frac{7}{11}$ $\frac{9}{13}$ $\frac{13}{17}$.
  Show clearly how you arrived at your answer.

7.3   Addition and subtraction of decimals

1   Work these out.
(a) $3 + 1.6$ (b) $4.2 - 1.1$ (c) $4.2 - 2.5$
(d) $7.2 + 3 + 0.8$ (e) $13.3 - 2.9$ (f) $14 - 5.2$

2   Copy these out and move the decimal point to make the calculations correct.
(a) $2.7 + 7.3 = 100$ (b) $0.7 + 0.3 = 10$ (c) $100 - 5.5 = 4.5$
(d) $6.8 - 155 = 5.25$ (e) $7.75 - 15 = 6.25$ (f) $100 - 9.5 = 0.5$

3   In a magic square, the numbers in each row,
column and diagonal add up to the same number.

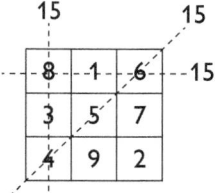

Copy and complete these magic squares.

(a)
0.4		0.8
	0.5	
0.2	0.7	

(b)
	3.3	1.8
2.7	2.1	
2.4		

(c)
		0.7
	1.45	
2.2		1.7

4   (a) Write these numbers in order of size, smallest first.
         0.56   0.567   0.088   0.525
(b) Which two numbers have the largest difference?
       Calculate the largest difference.
(c) Which two numbers have the smallest difference?
       Calculate the smallest difference.

5   Here is someone's homework. All the answers are wrong!
What mistakes do you think the student has made?
Calculate the correct answers.
(a) $5 + 2.1 = 2.6$ ✗   (b) $0.8 + 1.3 = 9.3$ ✗
(c) $25 - 2.2 = 0.3$ ✗   (d) $0.2 + 0.3 - 0.1 = 4$ ✗

6 **Investigation**

Write down the answer to $10\,000\,000 + 0.001$ (ten million add one-thousandth).
Now work out the answer with your calculator.
What do you notice? Has your calculator forgotten how to add?
Investigate with some other calculations like the one above.
Write two or three sentences explaining what is happening.

8 Co-ordinates

8.1 Co-ordinates

For this exercise you will need squared paper.

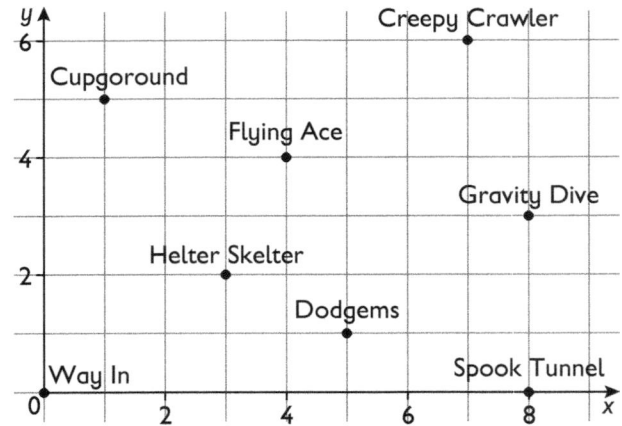

The grid shows some of the rides at a theme park.

1 Name the rides at:
(a) (3, 2)
(b) (8, 0)
(c) (1, 5).

2 Give the co-ordinates of:
(a) Flying Ace
(b) Dodgems
(c) Gravity Dive.

3 Which ride is furthest from the Way In?
Give its name and co-ordinates.

4 Draw a set of axes on squared paper.
Make the horizontal axis from 0 to 8 and the vertical axis from 0 to 6.
Mark these rides on your grid:
(a) Loopy at (1, 3)
(b) Highworld at (3, 6)
(c) Skyplunge at (0, 2)
(d) Swinghorse at (6, 3).

5 **Activity** ────────────────────────────────

For this activity, use a separate piece of squared paper.
(a) Draw your own co-ordinate map of a treasure island or a zoo.
(b) Below your map, write some questions about it, using co-ordinates.
On the back of the paper, write the answers to your questions.
(c) Swap your map with someone else and answer their questions.

8.2   Using co-ordinates in mathematics

For this exercise you will need squared paper.

1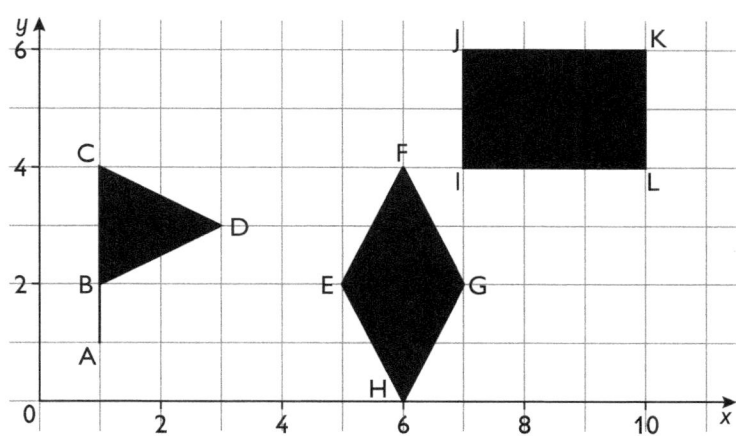

Write down the co-ordinates of points A to L.

2 Draw axes from 0 to 6 for x and y.
Plot: A(2, 2), B(6, 2), C(5, 4), D(3, 4).
Join ABCDA with ruled straight lines to make a trapezium.
  (a) Which angle is equal to angle C?
  (b) Which sides have equal length?
  (c) Which sides are parallel?

3 Draw axes from 0 to 6 for x and y.
Plot: A(3, 0), B(5, 3), C(3, 5).
Plot another point D so that ABCD is a kite.
Write down the co-ordinates of D.

4 Using centimetre squared paper, and a scale of 1 cm to 1 unit, draw a horizontal axis from 0 to 9 and a vertical axis from 0 to 6.
Plot these points and join them with ruled straight lines in this order.

A(0, 4), B(2, 6), C(4, 4), D(7, 4), E(9, 6), F(8, 4), G(8, 1), H(7, 1), I(7, 2), J(4, 2), K(4, 1), L(3, 1), M(3, 2), N(2, 4), O(1, 3)

Join the last point back to the first.
What have you drawn?

5 **Activity** ───

Make up your own co-ordinate picture and instructions to share with a friend.

8.3   Using negative numbers, fractions and decimals

For this exercise you will need squared paper and graph paper.

1 (a) Write down the co-ordinates of A and B.
Find the co-ordinates of the midpoint of AB.
(b) Find the midpoints of the lines CD and EF.

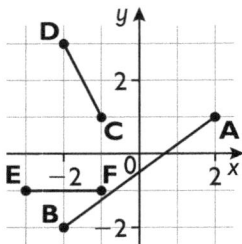

2 Draw axes from -3 to 3 for both x and y.
Plot the points A(1, 1), B(-3, -1), C(-2, -3) and join them in the order ABC.
ABCD is a rectangle. Find the co-ordinates of D.
Find the co-ordinates of the centre point of this rectangle.

3 Using 2 mm graph paper, draw axes from -5 to 5 for x and y.
(a) Draw the line joining A(2, 2) and B(-2, -3).
Write down the co-ordinates of the points where the line AB crosses the axes.
(b) Draw a line through A which is perpendicular to AB.
Write down the co-ordinates of the points where this line crosses the x and y axes.

4

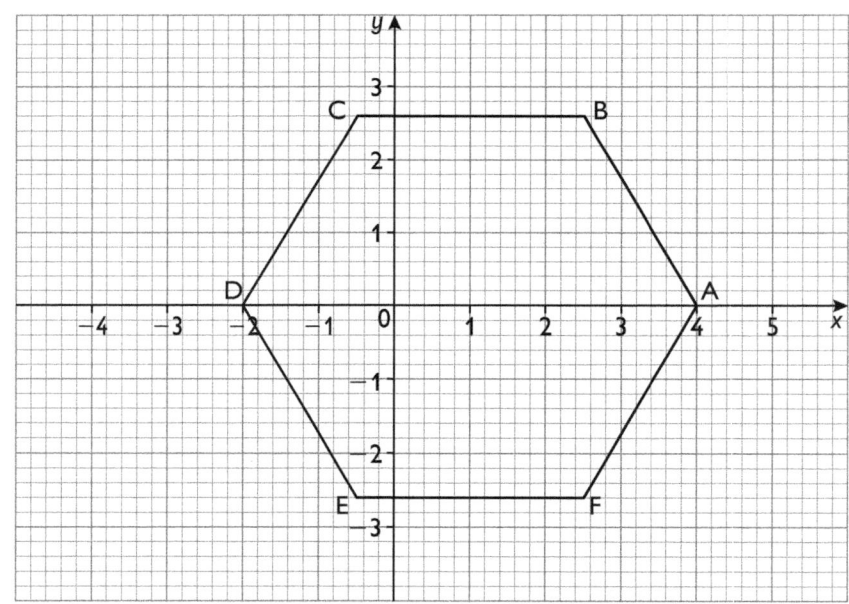

(a) Write down the co-ordinates of points A to F for this hexagon.
(b) Find the midpoints of AD, BE and CF.

5 Find the co-ordinates of the midpoints of the lines joining these points:
(a) (1, 0) and (7, -2)   (b) (2.4, -1) and (3.8, 5)   (c) (-4, 6.2) and (6, 20.8).

9 Fractions

9.1 Fractions

1 For each part, write down the fraction of the circle that is shaded.

(a) 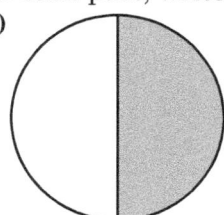 (b) 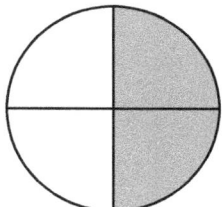 (c) 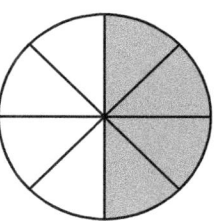 (d)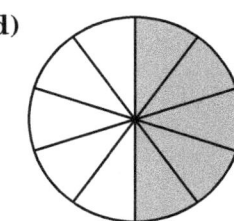

What do you notice?

2 Find: (a) $\frac{1}{2}$ of €64 (b) $\frac{1}{5}$ of 75 kg (c) $\frac{3}{4}$ of €56 (d) $\frac{2}{5}$ of 120 km

(e) $\frac{7}{10}$ of €370 (f) $\frac{5}{12}$ of 12 weeks (g) $\frac{1}{6}$ of 150 g (h) $\frac{7}{20}$ of 240 people

(i) $\frac{1}{8}$ of 96 cm (j) $\frac{2}{3}$ of 21 bananas.

3 How much is the deposit on these items?

(a)

Deposit = $\frac{1}{5}$ of price

(b) €400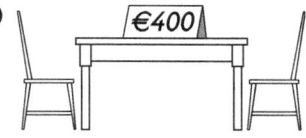

(c) €2000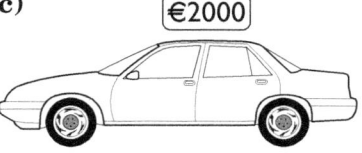

4 Which bottle contains more?

(a)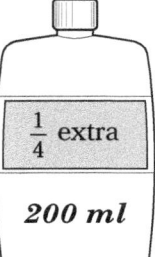

$\frac{1}{4}$ extra

200 ml

(b)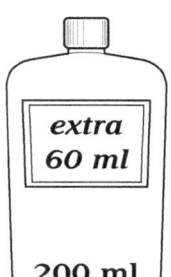

extra 60 ml

200 ml

5 In each part, find the missing numbers and answer the question that follows.

(a) $\dfrac{3}{4} = \dfrac{3 \times \square}{4 \times 6} = \dfrac{\square}{24}$

Is $\frac{3}{4}$ larger than $\frac{17}{24}$?

(b) $\dfrac{5}{7} = \dfrac{5 \times \square}{7 \times 6} = \dfrac{\square}{42}$

Is $\frac{5}{7}$ larger than $\frac{31}{42}$?

(c) In each case, write which of the pair is larger.

   (i) $\frac{2}{3}$ and $\frac{7}{12}$ (ii) $\frac{2}{5}$ and $\frac{11}{30}$ (iii) $\frac{5}{9}$ and $\frac{23}{36}$

6 In each case, write which is larger.

(a) $\frac{2}{3}$ and $\frac{3}{4}$ (b) $\frac{7}{10}$ and $\frac{18}{25}$

9.2   Multiplying fractions

In questions 1 to 4 work out the products, writing your answers as simply as possible.

1 (a) $\frac{1}{3} \times \frac{1}{5}$ (b) $\frac{1}{4} \times \frac{1}{7}$ (c) $\frac{3}{5} \times \frac{2}{5}$ (d) $\frac{3}{4} \times \frac{3}{8}$ (e) $\frac{1}{7} \times \frac{3}{5}$

  (f) $\frac{5}{6} \times \frac{4}{9}$ (g) $\frac{7}{11} \times \frac{5}{12}$ (h) $\frac{5}{8} \times \frac{7}{10}$ (i) $\frac{2}{11} \times \frac{4}{7}$ (j) $\frac{3}{10} \times \frac{2}{5}$

2 (a) $\frac{4}{5} \times \frac{3}{8}$ (b) $\frac{2}{5} \times \frac{3}{10}$ (c) $\frac{5}{6} \times \frac{1}{10}$ (d) $\frac{3}{4} \times \frac{2}{3}$ (e) $\frac{3}{8} \times \frac{4}{9}$

  (f) $\frac{5}{7} \times \frac{2}{15}$ (g) $\frac{6}{11} \times \frac{1}{12}$ (h) $\frac{2}{9} \times \frac{5}{6}$ (i) $\frac{3}{8} \times \frac{4}{7}$ (j) $\frac{3}{7} \times \frac{7}{12}$

3 (a) $\frac{4}{3} \times \frac{6}{5}$ (b) $\frac{7}{4} \times \frac{3}{7}$ (c) $\frac{3}{2} \times \frac{7}{6}$ (d) $\frac{3}{5} \times \frac{10}{9}$ (e) $\frac{10}{7} \times \frac{21}{8}$

  (f) $\frac{7}{2} \times \frac{5}{8}$ (g) $\frac{7}{5} \times \frac{5}{3}$ (h) $\frac{15}{8} \times \frac{11}{10}$ (i) $\frac{21}{10} \times \frac{4}{15}$ (j) $\frac{14}{9} \times \frac{14}{9}$

4 (a) $\frac{2}{5} \times 6$ (b) $\frac{5}{6} \times 12$ (c) $\frac{7}{10} \times 15$ (d) $\frac{8}{9} \times \frac{2}{3} \times \frac{9}{10}$ (e) $\frac{2}{7} \times 9$

  (f) $\frac{4}{5} \times \frac{5}{8} \times \frac{3}{7}$ (g) $\frac{6}{11} \times 5$ (h) $\frac{8}{3} \times \frac{4}{15} \times \frac{5}{16}$ (i) $\frac{3}{8} \times \frac{4}{15} \times \frac{7}{2}$ (j) $\frac{5}{2} \times 16$

9.3   Adding and subtracting fractions

1 Work out

  (a) $\frac{2}{5} + \frac{1}{5}$ (b) $\frac{3}{8} + \frac{5}{8}$ (c) $\frac{1}{3} + \frac{1}{4}$ (d) $\frac{2}{5} + \frac{1}{6}$

  (e) $\frac{3}{4} + \frac{1}{8}$ (f) $\frac{3}{16} + \frac{1}{4}$ (g) $\frac{3}{8} + \frac{2}{5}$ (h) $\frac{2}{9} + \frac{3}{4}$

2 (a) Find the smallest number which has the three numbers 3, 4 and 9 as factors.

  (b) Use the answer to (a) to work out

  $\frac{1}{3} + \frac{1}{4} + \frac{2}{9}$

3 Work out

  (a) $\frac{2}{3} - \frac{1}{4}$ (b) $\frac{3}{5} - \frac{1}{5}$ (c) $\frac{7}{9} - \frac{2}{3}$ (d) $\frac{5}{8} - \frac{5}{12}$

  (e) $\frac{9}{14} - \frac{2}{7}$ (f) $\frac{9}{10} - \frac{3}{4}$ (g) $\frac{5}{9} - \frac{2}{5}$ (h) $\frac{17}{20} - \frac{4}{5}$

4 Work out

  (a) $\frac{3}{8} + \frac{1}{4} + \frac{2}{5}$ (b) $\frac{5}{6} + \frac{2}{3} - \frac{3}{4}$ (c) $\frac{3}{20} + \frac{7}{10} - \frac{5}{8}$

5 Highborrow School fete raises money for the school and for charity. This year it gave $\frac{1}{3}$ to a medical charity and $\frac{1}{4}$ to a children's charity. What fraction is left for the school?

6 Sunny Holidays require an immediate deposit of $\frac{1}{3}$ and a further payment of $\frac{1}{5}$ later. What fraction of the cost of the holiday remains to be paid?

7 Sylvia walked $\frac{3}{8}$ of a mile, took a rest and then walked a further $\frac{2}{5}$ of a mile. How much has she walked so far?

8 Catley village had already raised $\frac{5}{12}$ towards rebuilding the village hall. After the summer fete it had raised $\frac{5}{8}$ of the money needed. What fraction did the summer fete raise?

9.4   Introduction to percentages

1 Write these percentages as fractions and cancel to their lowest form where possible.
   (a) 47%   **(b)** 30%   **(c)** 3%   **(d)** 64%   **(e)** 8%

2 Find 10% of each of these amounts.
   (a) €200   **(b)** 640 km   **(c)** €30   **(d)** 450 kg

3 There is a 20% reduction on goods in a sale. Find the money taken off these items:
   (a) a dress priced at €40   **(b)** a car priced at €4000
   (c) a TV priced at €450   **(d)** a painting priced at €600.

4 Which of these reductions is the greater and by how much?
   30% of €540 or 40% of €400

5 Find these quantities.
   (a) 50% of €640   **(b)** 80% of 300 kg   **(c)** 35% of €400   **(d)** 4% of 200 m

6 Raul got 78% in a test. What percentage did he get wrong?

7 15% of the price of a new car is profit.
   (a) What is the profit on a car priced at €7840?
   (b) What percentage of the price is the cost to build the car?

9.5   Ratio and proportion

1 In a drink of orange juice the ratio of orange concentrate to water is 1 : 5.
   (a) How much water is needed when:
      (i) 25 ml of orange concentrate is used   **(ii)** 120 ml of orange concentrate is used?
   (b) How much orange concentrate is needed when:
      (i) 400 ml of water is used   **(ii)** 60 ml of water is used?

2 Simplify these ratios.
   (a) 6 : 9   **(b)** 14 : 7   **(c)** 8 : 20   **(d)** 5 : 25
   (e) 12 : 16 : 40   **(f)** 56 : 14 : 28   **(g)** 40 : 24   **(h)** 3 : 9 : 24

3 Sparkle car wash is mixed with water in the ratio 1 : 6.
   (a) How much water is needed with 35 ml of Sparkle?
   (b) How much Sparkle is needed with 3 l of water?

4 A high-speed train has 120 first class seats and 300 standard class seats.
   (a) Write the ratio first class seats : standard class seats and simplify it as far as possible.
   (b) Half of the first class seats and two-thirds of the standard seats are occupied.
       What fraction of the seats in the train are occupied?

5 Fryers School has 450 boys and 600 girls.
   (a) Write down the ratio boys : girls in its simplest form.
   (b) On one day, one-fifth of the boys and one-tenth of the girls are absent. What fraction
       of the pupils are absent on that day?

10 Number patterns

10.1 Factors and multiples

1 Write the first four multiples of:
   (a) 2   **(b)** 5   **(c)** 7   **(d)** 10   **(e)** 12.

2 Which of the following numbers is not a multiple of 4?
   20   26   32   38   44   62   76

3 **(a)** Write the first ten multiples of 3.
   (b) Under each of your answers in **(a)** write the sum of the digits.
   (c) Write a rule for testing whether or not a number is divisible by 3.
   (d) Is 214 650 divisible by 3? Write some more numbers that are divisible by 3.
   (e) Using the same method, find a rule for testing whether or not a number is divisible
   by 9.

4 Write the multiples of 8 which are less than 50.
   Write the multiples of 6 which are less than 50.
   Write the smallest multiple that is common to both lists. This number is known as the
   Lowest Common Multiple, or LCM, of 8 and 6.

5 Two lights start flashing at the same time.
   One light flashes every 6 seconds and the other flashes every 9 seconds.
   After how long will the two lights flash together again?
   After how long will the two lights flash together for a third time?

6 Copy and complete this diagram to show the factors of 24.

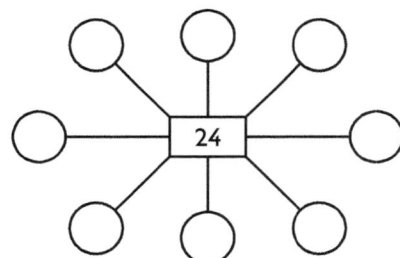

7 List all the factors of 32.

8 Which of the numbers 2, 3, 4, 5, 8 are factors of:
   (a) 30   **(b)** 72?

9 **(a)** List all the factors of 28 and 42.
   (b) What is the largest factor common to both of the lists? This number is called the
   Highest Common Factor, or HCF, of 28 and 42.

10 Which of these statements is true and which is false?
   Explain how you arrive at your answer.
   (a) The LCM of 2 and 6 is the same as the LCM of 3 and 6.
   (b) The multiples of 3 end in 1, 2, 3, 4, 6, 8 or 9.
   (c) The LCM of 6 and 9 is the same as the HCF of 36 and 54.

10.2 Squares

1 Copy and complete each table up to row 10. Make sure you follow the patterns.

1	1	1
2	$1 + 3$	4
3	$1 + 3 + 5$	9
4	$1 + 3 + 5 + 7$	⋮
⋮	⋮	⋮
10	⋮	⋮

1	1	1
2	$1 + 2 + 1$	⋮
3	$1 + 2 + 3 + 2 + 1$	⋮
4	⋮	⋮
⋮	⋮	⋮
10	⋮	⋮

The numbers in the right-hand column of each of these tables are called square numbers.
Explain why.

2 Write down the first ten square numbers.
Work out the difference between each consecutive pair of square numbers.
What do you notice about these numbers?

3 Use the table of squares to answer the following.

Number	Square
11	121
12	144
13	169
14	196
15	225
16	256
17	289
18	324
19	361
20	400
21	441
22	484
23	529
24	576
25	625
26	676

(a) 20^2 (b) 13^2 (c) 19^2
(d) 18^2 (e) 22^2 (f) 26^2
(g) $\sqrt{225}$ (h) $\sqrt{625}$ (i) $\sqrt{256}$
(j) $\sqrt{441}$ (k) $\sqrt{576}$

4 Use the table of squares to help you answer the following.
(a) $\sqrt{15^2 + 20^2}$ (b) $\sqrt{10^2 + 24^2}$
(c) $\sqrt{17^2 - 15^2}$ (d) $\sqrt{20^2 - 16^2}$

5 $\left(\frac{3}{4}\right)^2 = \frac{9}{16}$. Explain why.
Now, work out the following using the same method.
(a) $\left(\frac{1}{2}\right)^2$ (b) $\left(\frac{5}{8}\right)^2$ (c) $\left(\frac{9}{10}\right)^2$

6 Find
(a) 0.5^2 (b) 0.8^2 (c) 0.2^2

7 Find
(a) $\sqrt{64}$ (b) $\sqrt{16}$ (c) $\sqrt{1}$
(d) $\sqrt{\frac{4}{25}}$ (e) $\sqrt{\frac{9}{100}}$ (f) $\sqrt{\frac{49}{144}}$

8 Between which two whole numbers do the following lie?
(a) $\sqrt{20}$ (b) $\sqrt{150}$

10.3 Primes

1 (a) Write down all the prime numbers between 1 and 30.
   (b) Jamie says that when you add two prime numbers together the answer cannot be prime. Find some pairs of prime numbers to show that Jamie is wrong.

2 (a) Can two consecutive numbers both be prime?
   (b) Can three consecutive numbers all be prime?

3 Two prime numbers are multiplied together.
   When is the answer:
   (a) an even number
   (b) an odd number
   (c) a prime number?

4 Find two prime numbers that add to 100.
   How many different pairs of prime numbers add to 100?

5 Explain why two prime numbers can never add to 101.

6 This is a prime factor tree for 60.

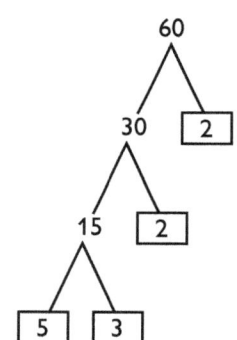

So 60 can be written as the product of prime factors $2 \times 2 \times 3 \times 5$.

Write the following numbers as a product of prime factors.
   (a) 12   (b) 100   (c) 78
   (d) 242   (e) 357

10.4 Cubes

1	2	3	4	5	6	7	.	.	.
2	4	6	8	10	12	14	.	.	.
3	6	9	12	15	18	21	.	.	.
4	8	12	16	20	24	28	.	.	.
.	.	.	.	.	.	.	.	.	.
.	.	.	.	.	.	.	.	.	.
.	.	.	.	.	.	.	.	.	.

(a) **(i)** Copy the above grid.

  (ii) Describe what the grid of numbers is.

  (iii) Extend the numbers in the grid by three more rows and three more columns.
    Draw in the lines.

  (iv) Find the total of each set of numbers between the lines.

  (v) Write the totals in order.

  (vi) Check that the totals are the cube numbers.

(b) The third set of numbers in the grid can be written

$$3 + 6 + 9 + 6 + 3$$

or

$$3(1 + 2 + 3 + 2 + 1).$$

  (i) Write the fourth and fifth sets of numbers in the same way.

  (ii) Without extending the grid, write the tenth set in this way.
    Find the total and check that the total equals 10^3.

2 Work out the cube root of each of these numbers.

  (a) 27   **(b)** 125

  (c) 216   **(d)** 512

3 Find the value of the following.

  (a) $\left(\frac{1}{4}\right)^3$ **(b)** $\left(\frac{2}{5}\right)^3$ **(c)** $\left(\frac{3}{8}\right)^3$

  (d) $\left(\frac{7}{10}\right)^3$ **(e)** $\left(\frac{1}{20}\right)^3$

4 Find the value of the following.

  (a) 0.5^3 **(b)** 0.6^3

  (c) 0.3^3 **(d)** 0.1^3

10.5 Powers of 2

1 An old legend tells how a young knight rescues a princess from an evil baron and returns her to her father, the king.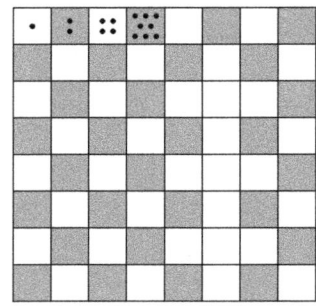
For bringing his daughter back safely, the king allows the knight to choose his own reward.
The knight just asks for some grains of corn.
'Look at this chess board' he says. 'Put one grain of corn on the first square, two on the second square, four on the third square, eight on the fourth square and double each time for all 64 squares.'
The king agrees, thinking that this is a small reward to ask.

(a) How many grains of corn will there be on
   (i)   the fifth square
   (ii) the sixth square
   (iii) the seventh square
   (iv) the tenth square?

(b) Use your calculator to find the number of grains on
   (i)   the twentieth square
   (ii) the fortieth square
   (iii) the sixty fourth square.

2 'As I was going to St Ives,
I met a man with seven wives.
Each wife had seven sacks.
Each sack had seven cats.
Each cat had seven kits.
Kits, cats, sacks and wives, how many were going to St Ives?'

You may have seen this riddle before. There is trick in the question and the answer is one person only (I) was *going* to St Ives.
But how many were coming *from* St Ives?

11 Everyday measures

11.1 Length

1 What metric unit would you use to measure:
 (a) the width of this book
 (b) the length of a drawing pin
 (c) the distance from London to Paris
 (d) the distance around a running track?

2 Match each of the distances in list A with *two* appropriate measures of length from list B.

List A	List B	
The distance from Cardiff to Chester	1 inch	55 yards
The width of a football pitch	280 mm	69 cm
The length of a man's foot	150 miles	25 mm
The height of a classroom	50 m	8 ft 3 inches
The width of a €2 piece	250 cm	27 inches
The circumference of a football	11 inches	240 km

3 Measure each of the following lines **(a)** to the nearest millimetre and **(b)** to the nearest centimetre.

4 Change these lengths to millimetres.
 (a) 7 m   **(b)** $\frac{1}{10}$ m   **(c)** 15.4 cm   **(d)** 0.35 m

5 Change these lengths to metres.
 (a) 160 cm   **(b)** $\frac{1}{100}$ km   **(c)** 160 mm   **(d)** 0.24 km

6 This sketch map shows the position of Sandra's house and each of her friends' houses. All distances are in metres.
 (a) One day Sandra decides to visit her four friends.
 Which route should she take in order to walk the shortest distance?
 (b) What is the shortest distance?
 (c) Write the shortest distance in kilometres.

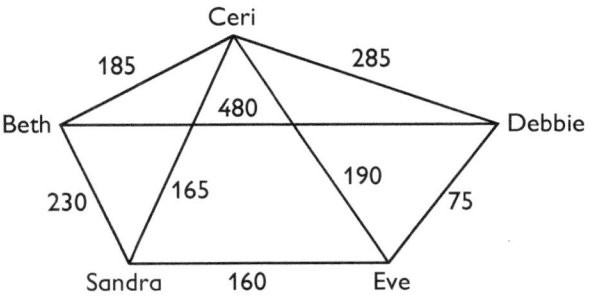

11.2 Mass (or weight)

1 Change these masses into kilograms.
(a) 4000 g   (b) 1600 g   (c) 75 g   (d) 940 g

2 Change these masses into grams.
(a) 6 kg   (b) 3.5 kg   (c) 0.64 kg   (d) $\frac{1}{10}$ kg

3 Howard opens a 1 kg bag of sugar.
He uses 350 grams to make some cakes.
How much sugar is left in the bag?

4 Add up these five masses.

123 kg,   850 g,   3 kg,   1030 g,   75 g

5 These are the ingredients needed
to make muesli.
Find the total weight of the ingredients:
(a) (i)   in grams
     (ii) in kilograms
(b) in Imperial units.

Muesli		
Wholewheat flakes	225 g	(8 oz)
Oat flakes	225 g	(8 oz)
Barley flakes	225 g	(8 oz)
Mixed roasted nuts	350 g	(12 oz)
Sunflower seeds	175 g	(6 oz)
Sultanas	450 g	(16 oz)
Raisins	225 g	(8 oz)
Dried prunes	100 g	(4 oz)

6 Asif weighs 71 kg. Imran weighs 68 000 g.
Who weighs more, Asif or Imran, and by how much?

7 A jug can hold 250 ml. How many times can the jug be filled from a bottle containing
2 litres?

8 Trevor makes some salad dressing.
He mixes 200 ml of lemon juice and 150 ml of orange juice with 0.5 litres of olive oil.
How much dressing does he make altogether?
Give your answer   (a) in millilitres   (b) in litres.

11.3 Time

1 Write these times using am or pm.
(a) 09:15   (b) 21:15   (c) 17:05   (d) 10:42

2 Write these times using the 24-hour clock.
(a) 7.30 am   (b) 7.30 pm   (c) ten to one in the morning
(d) half past six in the evening   (e) three thirty in the afternoon

3 (a) How many minutes in 7 hours?
(b) How many minutes in 480 seconds?

4 Here is part of the timetable for lessons at one school.
(a) How long does Registration and Assembly last?
(b) Lesson 2 last $1\frac{1}{4}$ hours. At what time does it finish?
(c) Lunch last 50 minutes.
Lesson 4 lasts $1\frac{1}{4}$ hours.
Complete the timetable for Lunch and Lesson 4.

Activity	Time of day
Registration & Assembly	08:35–09:05
Lesson 1	09:05–10:15
Break	10:15–10:35
Lesson 2	10:35–
Lesson 3	–13:05
Lunch	13:05–
Lesson 4	–

5 Aled travels to his office each day.
(a) On Tuesday he left home at 8.45 am and arrived in his office at 9.20 am.
How long did the journey take?
(b) He left the office for home at 5.50 pm.
The journey took 45 minutes.
At what time did he return home?

6 This is part of a timetable for a bus company.

Cardiff Central	08:50	09:20	09:55	10:55	12:05
Park Place	08:59	09:29	10:04	11:04	12:14
Whitchurch	09:16	09:46	–	–	12:31
Pontypridd	09:38	10:08	10:35	11:38	12:53
Abercynon	10:02	–	10:57	–	13:17
Mountain Ash	10:13	10:38	11:06	12:08	13:28
Aberdare	10:15	10:50	11:18	12:20	13:40

(a) Cerys catches the bus that leaves Cardiff Central at 09:20.
   (i) At what time does the bus arrive in Aberdare?
   (ii) How long does the journey take?
(b) Which bus takes the least time from Cardiff to Aberdare?
(c) Between which two stops is the bus at 1 pm?
(d) Dipak needs to be in Aberdare by midday.
What is the time of the latest bus he can catch from Park Place?
(e) Trevor catches the bus that leaves Whitchurch at 12:31.
He has arranged a meeting in Abercynon for 1.45 pm.
How long after arriving in Abercynon is his meeting due to start?

12 Flat shapes

12.1 Flat shapes

For this exercise you will need to know the names triangle, quadrilateral, pentagon and hexagon.

1 Name these shapes.
Say whether they are regular or irregular.

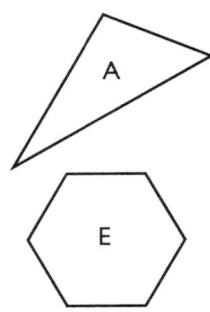 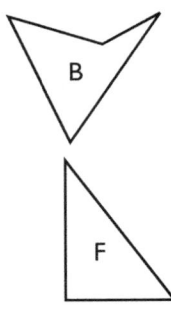 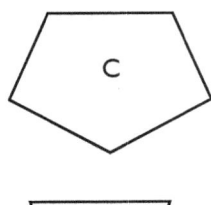 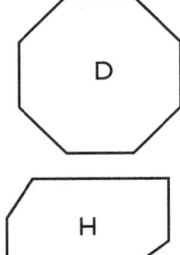 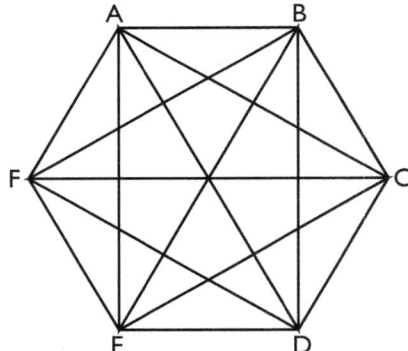

2 Draw the following shapes:
  (a) a regular triangle
  (b) a regular pentagon
  (c) a regular quadrilateral
  (d) an irregular triangle with one axis of symmetry
  (e) an irregular hexagon
  (f) an irregular quadrilateral with two axes of symmetry.

3 **Investigation**

The diagram shows a regular hexagon with all its diagonals.
The diagonals divide the hexagon into other shapes, for example BCEF is a rectangle.
Write down two sets of letters which describe:
  (a) a triangle   **(b)** a rectangle   **(c)** a pentagon   **(d)** a quadrilateral.

12.2 Classifying triangles and quadrilaterals

1 Write the special names of each of the following shapes.

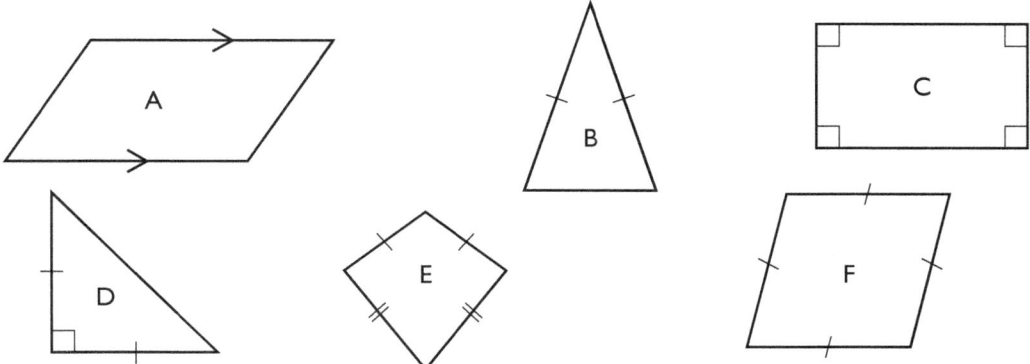

2 Describe the symmetries (if any) of the shapes in question 1.

3 Draw the following shapes:
   (a) a trapezium with one axis of symmetry
   (b) an arrowhead
   (c) a scalene triangle
   (d) a quadrilateral with four axes of symmetry.

4 State which of the following are true:
   (a) all squares are rectangles
   (b) some rhombuses are squares
   (c) all triangles have one axis of symmetry
   (d) all rhombuses are parallelograms
   (e) all rectangles are parallelograms
   (f) all kites are quadrilaterals
   (g) all parallelograms are trapeziums.

5 **Investigation**

How many parallelograms are there in this diagram?

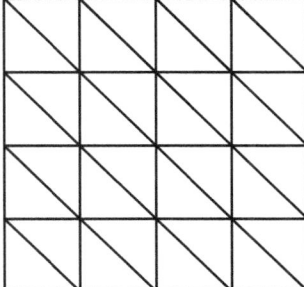

12.3 Circles

For this exercise you will need a pair of compasses, a sharp pencil and a ruler.

1 What is the diameter of a circle with radius:
    (a) 6 cm   **(b)** 4 cm   **(c)** 2.5 cm?

2 What is the radius of a circle with diameter:
    (a) 10 cm   **(b)** 9 cm   **(c)** 14 cm?

3 **(a)** Open your compasses to about 5 cm and draw a circle.
    (b) Without adjusting your compasses use them to draw this pattern.

4 Use your compasses to draw this shape. It is made from semi-circles.

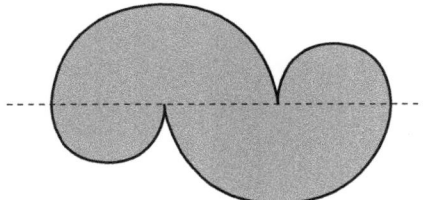

5 **Investigation**

The diagram shows a horse in a field. It is tethered to a corner of its stable.
Illustrate the region in which the horse can graze.

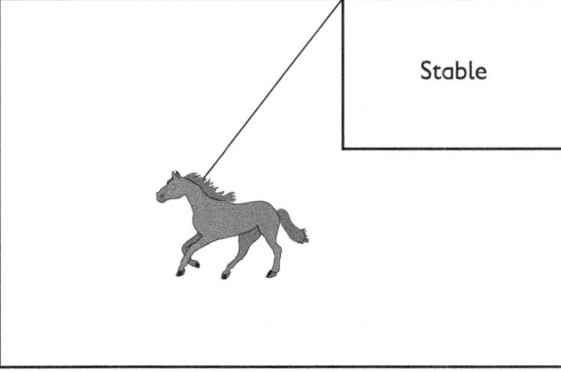

12.4 Constructing triangles

For this exercise you will need a pair of compasses, a protractor and a ruler to draw accurate triangles. Remember you should not rub out the arcs and 'extra bits' of line as they are an important part of the construction. You will need a protractor to measure the angles.

1 Construct:
  (a) an equilateral triangle with sides of 4 cm
  (b) an isosceles triangle with sides 8 cm, 5 cm and 5 cm
  (c) a scalene triangle with sides 6 cm, 8 cm and 10 cm.
  Measure and write the size of the angles in your triangles.

2 Without actually drawing them write whether or not you can construct triangles with the following sides.
  (a) 10 cm, 5 cm, 4 cm
  (b) 5 cm, 4 cm, 6 cm
  (c) 12 cm, 6 cm, 6 cm

3 Construct an equilateral triangle with sides of 8 cm.
  Mark the middle point of one side and join it to the opposite vertex as in the diagram.
  This line is called a median.
  Draw the other two medians.
  If your diagram is accurate these lines should all meet at a point.

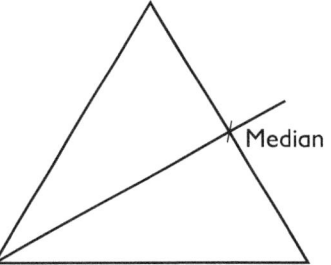

4 **Investigation**

This part of a tile pattern is made with triangular tiles.
These tiles are all equilateral triangles.
Angela says: 'Any triangle can be used to make a tile pattern.'
Investigate Angela's statement with isosceles and scalene triangles.
Your tiles should all be identical.
Draw and colour a pattern using your own triangular tile.

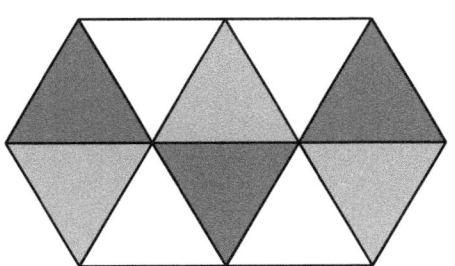

12.5 Constructions with angles

For this exercise you will need a pair of compasses, a protractor and a ruler to construct triangles with given sides and angles.

1 Make accurate constructions of these triangles.
For each one measure and write the length of the other two sides.

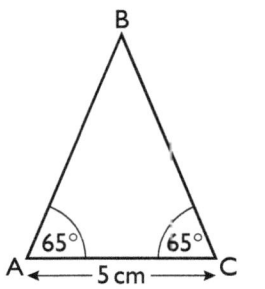
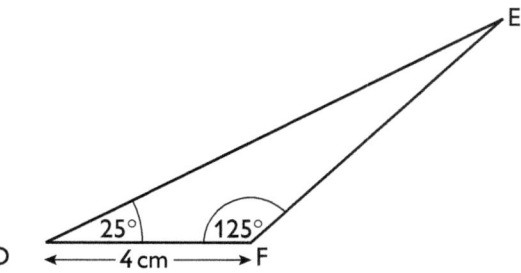

2 Use ruler and compasses only to construct this triangle.
Measure and write the length of the other two sides.

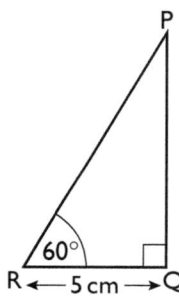

3 Draw a triangle with sides of length 5 cm and 6 cm with an angle of 52° between them.
Measure the other side and the remaining angles.

4 **Investigation**

Parveen was given these instructions.
Draw a triangle ABC with AB = 9 cm, angle
ABC = 32° and AC = 6 cm.
Here is Parveen's construction.
Explain what he has done.
How many possible triangles are there?

Investigate these instructions and say whether
you can draw 0, 1 or 2 triangles.
(a) AB = 9.5 cm, AC = 8 cm, angle ABC = 42°
(b) AB = 7 cm, AC = 9.2 cm, angle ABC = 100°
(c) AB = 13 cm, AC = 9 cm, angle ABC = 65°
(d) AB = 10 cm, AC = 12 cm, angle ABC = 41°

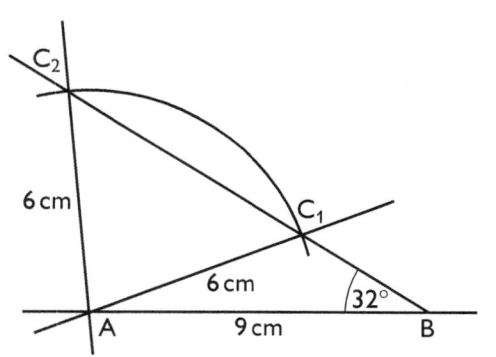

13 Multiplying and dividing decimals

13.1 Multiplying decimals

1 Work out the total cost of these items:
(a) five cinema tickets at €3.25 each
(b) eight rolls of wallpaper at €5.79 each
(c) two tins of paint at €14.78 each
(d) three boxes of chocolates at €3.95 each.

2 One day the exchange rate for pounds is €1 = £0.69.
(a) Change these amounts into pounds:
    (i) €4   (ii) €7   (iii) €15.
(b) Sally buys a camera for a holiday in England. She pays €95.
    She sees the same camera on sale in England for £62.
    How much cheaper is the camera in England?

3 Work out
(a) 0.6×0.4 (b) 0.03×0.9 (c) 0.08×0.3
(d) 0.7×0.4 (e) 0.5×0.3 (f) 0.03×0.06
(g) 0.006×0.6 (h) 0.09^2 (i) 0.02^2
(j) 0.5×0.003 (k) 0.009×0.2 (l) 0.07×0.007

4 Work out
(a) 12.6×0.4 (b) 1.03×9 (c) 6.8×0.7
(d) 7.14×0.5 (e) 32.5×0.3 (f) 13.2×0.57
(g) 16.1×0.14 (h) 8.79×0.22 (j) 0.25^2
(j) 6.54×0.32 (k) 0.096×0.8 (l) 3.37×0.11

5 Copy and complete these calculations by putting the decimal point in the correct place.
(a) $3.6 \times 4 = 144$ (b) $6.5 \times 3 = 195$
(c) $5.7 \times 0.8 = 456$ (d) $32.4 \times 0.07 = 2268$
(e) $0.07 \times 0.34 = 238$ (f) $7.34 \times 1.29 = 94\,686$
(g) $0.89 \times 6.45 = 57\,405$ (h) $11.87 \times 8 = 9496$
(i) $0.85 \times 0.6 = 51$ (j) $13.8 \times 21.6 = 29\,808$

6 A cuboid measures 8.6 cm long, 5.5 cm wide and 3.2 cm high.
(a) What are these measurements in mm?
(b) Calculate the volume of the cuboid in mm^3.
(c) $1\,cm^3 = 1000\,mm^3$.
    Use this to work out $8.6 \times 5.5 \times 3.2$.

7 A theme park charges €12.75 for adults and €8.25 for children. Mr and Mrs Tomlinson
and their three children visit the park for a day.
Work out the total cost for the family.

8 A toll bridge charges €3.75 for a return trip across the bridge.
Asif crosses the bridge each day to go to work and to return home.
He makes 25 return trips across the bridge in a month.
Asif buys a monthly season ticket. It costs €88.
How much does Asif save?

9 **(a)** Multiply
- **(i)** 2.6×0.3 **(ii)** 2.6×0.9 **(iii)** 2.6×0.4
- **(iv)** 9.1×0.02 **(v)** 9.1×0.3 **(vi)** 9.1×0.07
- **(vii)** 6.4×0.004 **(viii)** 6.4×0.2 **(ix)** 6.4×0.05

(b) Complete this sentence:
'When a number is multiplied by a number which is less than 1, it _____.'

10 **(a)** Multiply
- **(i)** 4.1×5 **(ii)** 4.1×9 **(iii)** 4.1×6
- **(iv)** 3.81×10 **(v)** 3.81×7 **(vi)** 3.81×2
- **(vii)** 12.3×3 **(viii)** 12.3×2 **(ix)** 12.3×8

(b) Complete this sentence:
'When a number is multiplied by a number which is bigger than 1, it _____.'

13.2 Division of decimals by a whole number

1 A pack of six chocolate bars costs €1.44. What is the cost of one bar?

2 A pack of five mini discs costs €4.95. What is the cost of one mini disc?

3 Divide each of these numbers by 5.
- **(a)** 3.6   **(b)** 4.3   **(c)** 1   **(d)** 12.8   **(e)** 0.52   **(f)** 7.4

4 Divide each of these numbers by 8.
- **(a)** 5.8   **(b)** 3.2   **(c)** 3   **(d)** 23.6   **(e)** 3.22   **(f)** 9.7

5 Divide each of these numbers by 9.
- **(a)** 3.6   **(b)** 3.24   **(c)** 23   **(d)** 7   **(e)** 4.8   **(f)** 15.8

6 Work out
- **(a)** $8 \div 10$ **(b)** $0.8 \div 10$ **(c)** $0.008 \div 10$ **(d)** $8 \div 100$
- **(e)** $0.8 \div 100$ **(f)** $0.008 \div 1000$ **(g)** $0.08 \div 1000$ **(h)** $8 \div 10\,000$

7 A company sells cans of cola in packs.
A pack of four costs €1.16 and a pack of six costs €1.92.
Which pack is cheaper per can and by how much?

8 For each of these numbers **(i)** multiply it by 50 and **(ii)** divide it by 2.
- **(a)** 6   **(b)** 3.2   **(c)** 0.8   **(d)** 8.13   **(e)** 0.46

9 This is a copy of Sandra's till receipt.
Howard visits the same supermarket
and buys 1 jar of strawberry jam,
3 Granny Smith apples, six eggs, a 2 litre
bottle of milk, 2 packets of plain crisps
and 2 bars of chocolate.
What does Howard's bill come to?

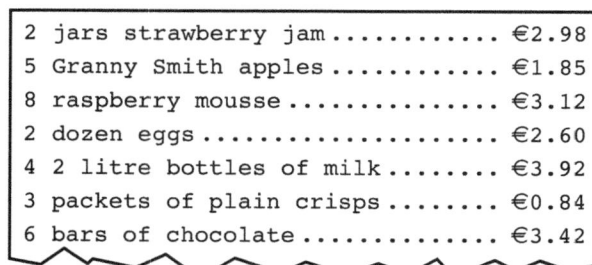

```
2 jars strawberry jam ........... €2.98
5 Granny Smith apples ........... €1.85
8 raspberry mousse .............. €3.12
2 dozen eggs .................... €2.60
4 2 litre bottles of milk ....... €3.92
3 packets of plain crisps ....... €0.84
6 bars of chocolate ............. €3.42
```

10 A supermarket sells a tube of chocolate drops for 22cent.
For one weekend they have this special offer 'BUY TWO TUBES, GET ONE FREE'.
Work out the maximum number of tubes Colin can buy for €2.

13.3 Division of decimals (short division)

1 (a) How many 5cent pieces are there in:
   (i) €3   (ii) €4   (iii) €7   (iv) €10   (v) €8.50?
   (b) Divide these numbers by 0.05.
   (i) 3.00   (ii) 4.00   (iii) 7.00   (iv) 10.00   (v) 8.50

2 (a) How many quarter-hours are there in:
   (i) 2 hours   (ii) 3 hours   (iii) 5 hours   (iv) $6\frac{1}{2}$ hours   (v) $4\frac{1}{4}$ hours?
   (b) Divide these numbers by 0.25.
   (i) 2   (ii) 3   (iii) 5   (iv) 6.5   (v) 4.25

3 Work out
   (a) $4.5 \div 0.3$ (b) $9.36 \div 0.4$ (c) $3.85 \div 0.5$
   (d) $7.105 \div 0.7$ (e) $5.112 \div 0.9$ (f) $6.304 \div 0.8$

4 Work out
   (a) $3.15 \div 0.07$ (b) $0.336 \div 0.04$ (c) $25.11 \div 0.09$
   (d) $34.05 \div 0.06$ (e) $47.825 \div 0.05$ (f) $0.0264 \div 0.03$

5 (a) Divide
   (i) $90.5 \div 5$ (ii) $34.6 \div 10$ (iii) $14.1 \div 6$
   (iv) $43.83 \div 9$ (v) $22.82 \div 7$ (vi) $608.8 \div 2$
   (vii) $42.3 \div 3$ (viii) $12.16 \div 4$ (ix) $55.02 \div 7$
   (b) Complete this sentence:
   'When a number is divided by a number which is greater than 1, it _____.'

6 A packet of biscuits is 25.6 cm long. Each biscuit is 0.8 cm thick.
   How many biscuits are in the packet?

7 A shop sells tablemats. On the counter it has a pile 33.2 cm high.
   Each tablemat is 0.4 cm thick.
   How many tablemats are in the pile?

8 Work out
   (a) $0.92 \div 0.08$ (b) $2.553 \div 0.006$ (c) $50.1 \div 0.05$
   (d) $0.345 \div 0.001$ (e) $7.85 \div 0.02$ (f) $0.0261 \div 0.003$
   (g) $13.5 \div 0.09$ (h) $8.115 \div 0.005$ (i) $2.5 \div 0.008$
   (j) $18.05 \div 0.002$ (k) $19.87 \div 0.1$ (l) $10.032 \div 0.004$

13.4 Equivalent percentages, fractions and decimals

1 Change these percentages to fractions and cancel to their lowest terms.
  (a) 20%   **(b)** 70%   **(c)** 15%

2 Change these percentages to fractions in their lowest terms.
  (a) 60% increase in car sales.
  (b) 5% of workers laid off.
  (c) 34% of cheese is fat.
  (d) 80% of pupils in a school eat school lunch.

3 Change these percentages to decimals.
  (a) 20%   **(b)** 25%   **(c)** 70%   **(d)** 44%
  (e) 8%   **(f)** 13%   **(g)** 95%   **(h)** $17\frac{1}{2}\%$

4 Change these decimals to percentages.
  (a) 0.35   **(b)** 0.5   **(c)** 0.05   **(d)** 0.19
  (e) 0.65   **(f)** 0.44   **(g)** 0.72   **(h)** 0.01

5 Change these fractions to percentages.
  (a) $\frac{2}{5}$ **(b)** $\frac{3}{10}$ **(c)** $\frac{11}{20}$ **(d)** $\frac{9}{50}$ **(e)** $\frac{8}{25}$

6 Change these decimals to fractions.
  (a) 0.3   **(b)** 0.5   **(c)** 0.6   **(d)** 0.18   **(e)** 0.36
  (f) 0.73   **(g)** 0.44   **(h)** 0.02   **(i)** 0.28   **(j)** 0.52

7 This is a record of Alan's exam results.
  English $\frac{57}{100}$ Science $\frac{32}{50}$ French $\frac{18}{30}$ History $\frac{11}{25}$
  Music $\frac{16}{40}$ Art $\frac{40}{50}$ Maths $\frac{13}{20}$
  Write these marks in order starting with the highest percentage.

8 **(a)** In a school there are 200 Year 7 students.
     On a Monday 24 students were absent.
     (i)   What percentage of the students were absent?
     (ii) What percentage of the students were not absent?
  (b) In the school 25 students study Maths in the sixth form.
     23 of the students are expected to pass their exam.
     (i)   What percentage are expected to pass their exam?
     (ii) What percentage are not expected to pass their exam?
  (c) There are 500 students in the lower school.
     350 of the students take part in a sponsored walk.
     (i)   What percentage take part in the walk?
     (ii) What percentage do not take part in the walk?

14 Number machines

14.1 Number machines

1 Work out the missing numbers in these machines.

(a)

3 → | × 3 | → | + 5 | → ?

(b)

2 → | × 6 | → | − 5 | → ?

(c)

5 → | × 2 | → | + 4 | → ?

(d)

8 → | × 5 | → | − 30 | → ?

(e)

3 → | × 2 | → | + 4 | → ?

(f)

9 → | × 3 | → | − 17 | → ?

2 Andy charges €10 for hanging a roll of wallpaper plus a basic charge of €80.
The cost of hiring Andy is calculated using this number machine.

No. of rolls → | × 10 | → | + 80 | → Hire cost in euro

Work out the cost of hiring Andy to hang:
(a) 5 rolls   **(b)** 8 rolls   **(c)** 13 rolls.

3 **(a)** Ben charges €12 for hanging a roll of wallpaper plus a basic charge of €70.
Copy and complete the number machine to show this information.

No. of rolls → | | → | | → Hire cost in euro

(b) Carla charges €15 per roll plus a basic charge of €30.
Draw a number machine to show this information.
(c) Would Ben or Carla charge more to hang 12 rolls of wallpaper and by how much?
(d) Ben charges a customer €178.
How many rolls of wallpaper did he hang?
(e) Carla charges a customer €120.
How many rolls of wallpaper did she hang?

4 **(a)** A florist sells flower arrangements in a basket.
She charges 50cent per flower plus €3 for the basket.
Draw a number machine to represent this information.
(b) These number machines represent the charges made by two other florists.

No. of flowers → | × 60cent | → | + €2 | → €14

No. of flowers → | × 40cent | → | + €5 | → €17

Work out the number of flowers by reversing each number machine. Check both your answers.

14.2 Finding the rule

1 Work out the rules for the following *adding* and *subtracting* number machines.

(a)
3 ⟶ ⬡ ⟶ 13

(b)
12 ⟶ ⬡ ⟶ 5

(c)
20 ⟶ ⬡ ⟶ 10

(d)
25 ⟶ ⬡ ⟶ 30

(e)
12 ⟶ ⬡ ⟶ 16

(f)
11 ⟶ ⬡ ⟶ 7

Write your rules in words.

2 Work out the rules for the following *multiplying* and *dividing* number machines.

(a)
40 ⟶ ⬡ ⟶ 4

(b)
28 ⟶ ⬡ ⟶ 4

(c)
3 ⟶ ⬡ ⟶ 12

(d)
5 ⟶ ⬡ ⟶ 20

(e)
18 ⟶ ⬡ ⟶ 2

(f)
7 ⟶ ⬡ ⟶ 1

Write your rules in words.

3 Look at these number machines.

(a)
4 ⟶ ⬡ ⟶ ⬡ ⟶ 7

(b)
5 ⟶ ⬡ ⟶ ⬡ ⟶ 12

(c)
3 ⟶ ⬡ ⟶ ⬡ ⟶ 8

(d)
12 ⟶ ⬡ ⟶ ⬡ ⟶ 10

Write three possible rules for each number machine.

4 Here are the contents of six boxes.

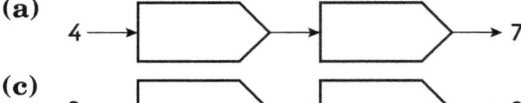 ⟨× 3⟩ ⟨− 8⟩ ⟨+ 9⟩ ⟨× 2⟩ ⟨− 4⟩ ⟨× 4⟩

Use each box once only to complete the chains below.

(a)
3 ⟶ ⬡ ⟶ 12

(b)
6 ⟶ ⬡ ⟶ ⬡ ⟶ 21

(c)
2 ⟶ ⬡ ⟶ ⬡ ⟶ 2

(d)
10 ⟶ ⬡ ⟶ 2

5 **(a)** Write three possible rules for this number machine.

4 ⟶ ⬡ ⟶ ⬡ ⟶ 17

This table shows some inputs and outputs for a number machine.

(b) What would be the output if the input was
(i) 5 **(ii)** 10 **(iii)** 30 **(iv)** 100?

(c) What is the rule for the number machine?

Input	1	2	3	4
Output	2	7	12	17

14.3 Writing down the rule

1 Sandra investigates how a canteen works out its charges for buffets.
The canteen charges €21 for a party of eight people.
It uses this number machine for calculating its charges.

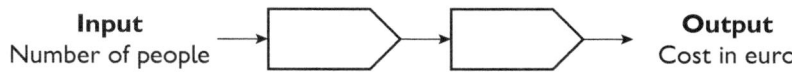

Input Number of people → → **Output** Cost in euro

(a) Write four rules that the canteen could be using.

Sandra collects some more data to help her
find the rule, as shown in the table.

(b) What is the cost when the number of
people is:
(i) 4   (ii) 7   (iii) 15   (iv) 20?

(c) How many people are in the party when
the cost is:
(i) €17   (ii) €27   (iii) €33   (iv) €65?

Number of people	Cost (in €)
3	11
5	15
8	21
10	25

(d) Write the rule that the canteen is using.
Use P to represent the number of people and C to represent the cost.

2 Huma is fixing posters to a notice board using pins.
The diagram shows how the posters overlap.

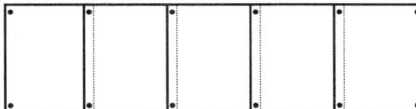

(a) Copy and complete this table.

Number of posters	1	2	3	4	5	6	7	8
Number of pins	4	6	8					

(b) Rewrite the information in the table in the form of a number machine.

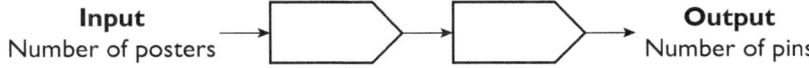

Input Number of posters → → **Output** Number of pins

(c) Write this rule using letters.
Represent the number of posters by n and the number of pins by p.

3 A company uses this table to work out costs for
hiring out concrete mixers.

(a) What is the cost of hiring a concrete mixer for
(i)   7 days
(ii) 10 days
(iii) 12 days?

(b) Write the rule that the company is using.
Use d to represent the number of days and c to
represent the cost.

Number of days	Cost (in €)
1	9
2	12
3	15
4	18

15 Scale

15.1 Scale models

1 Scale models are used to test the safety of buildings.
Wind tunnels are used, for example, to investigate the effects of wind gusts.

A 100 m tall building is to be tested using a 50 cm tall model.
(a) What is the scale of the model?
(b) The Empire State Building is 381 m tall.
Using this scale how tall would a model of it be?
(c) Using the same scale, about how tall should a model person be?

2 Scale models of boats are tested in water tanks before the actual boats are built.
A model is built to a 1 : 4 scale.

Complete this table showing various measurements on the model and on the 'real' boat.

Measurement	'Real' length	Length on model
Length	20 m	**(a)**
Width	**(b)**	1 m
Depth of hull under water	1 m	**(c)**

(d) The model boat has two boat propellers.
How many propellers should the actual boat have?
(e) One of the test tanks at the University of Southampton can test models up to 6 m
long. A planned super tanker is 300 m long.
What scale should be used for the model?
(f) Scale models of sailing boats can also be tested using large fans.
A rectangular sail on a 1 : 5 model sailing boat measures 60 cm by 80 cm.
(i) What are the dimensions of the 'real' sail?
(ii) What is the area of the model sail? (Don't forget the units.)
(iii) What is the area of the real sail? (Don't forget the units.)

3 Which of the objects in the table will go through this doorway?

Scale 1:50

	Wardrobe	Toy chest	Chest of draws
Height	210 cm	1 m	75 cm
Depth	50 cm	90 cm	60 cm
Width	110 cm	80 cm	90 cm

15.2 Drawing plans

1 Copy this table and fill in the missing numbers.
(Don't forget to use the correct units where necessary.)

Measurement		Scale used
On plan	In 'real-life'	
4 cm	100 cm	**(a)**
20 mm	10 m	**(b)**
(c)	5 cm	1 : 20
1 mm	**(d)**	1 : 40
2 cm	1.2 m	**(e)**

2 A notice board measures 230 cm by 130 cm.

Amy wants to stick five notices on the board.
Their dimensions are shown in the table below.

Make a scale drawing of the notice board.
Show how all the notices can be fitted onto it.

Use a scale of 1 cm to 20 cm.

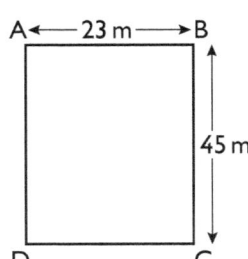

	Width (cm)	Length (cm)
Notice 1	80	40
Notice 2	70	65
Notice 3	70	75
Notice 4	70	120
Notice 5	60	60

3 This is a rough sketch of a rectangular field.
The field measures 45 m by 23 m.
(a) Make a scale drawing of the field.
Use a scale of 1 cm to 10 m.
(b) Use your scale drawing to find how much shorter it is to
walk along AC than along the sides AB and BC.

A ◄— 23 m —► B

45 m

D   C

15.3 Map scales

1 This is an old map of a small town.
Use a thread to help with measuring curves.
Measure to the nearest corner of buildings.

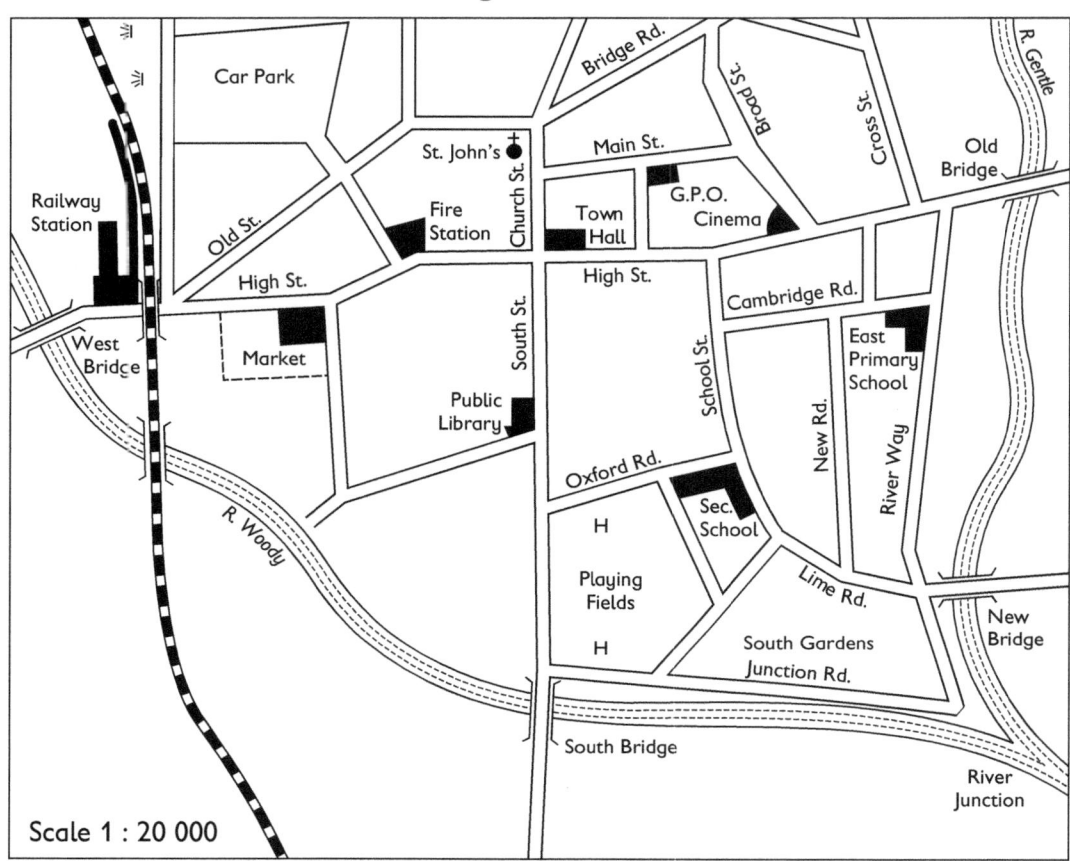

Scale 1 : 20 000

(a) Calculate these distances to the nearest 50 m:

    (i) from the station to the market

    (ii) from the town hall to the fire station

    (iii) from the secondary school to the cinema

    (iv) from South Bridge to Old Bridge by river.

(b) Plan a walk of about 5 km which begins and ends at the Library.
Check your answer by measuring.

2 Make scale drawings of these games fields.

    (a) A cricket square (30 m by 30 m). Use a scale of 1 : 1000.

    (b) A soccer pitch (120 m by 80 m). Use a scale of 1 : 2000.

3 Which map scale shows the greater detail:
1 inch to the mile or 1 cm to 1 kilometre?

You may find the information in this box useful.

> 12 inches make 1 foot
> 1 mile is 5280 feet

16 Averages

16.1 The mean

1 Find the mean of each of these sets of numbers.
(a) 5, 5, 5, 4, 1, 1, 7   (b) 3, 12, 7, 8, 15, 6
(c) 12, 9, 15, 6, 6, 13, 20, 14, 30, 17   (d) 21, 32, 19, 25

2 This table shows the number of hours of sunshine each day for a week in August for two seaside resorts.

	Sun	Mon	Tue	Wed	Thu	Fri	Sat
Resort A	2	4	8	11	9	10	5
Resort B	6	7	5	7	7	4	6

Work out the mean number of hours of sunshine for each resort.

3 Sabiha wants to compare the number of times the students in her class used their mobile phones each week. Here are her results.

Girls:	26	31	14	8	12	27	35	21	13
	8	20	14	25	12	17	19	16	15
Boys:	14	18	17	11	13	19	8	10	24
	23	14	18						

(a) Work out the mean number of times for (i) the girls (ii) the boys (iii) the whole class.
(b) Which group uses its mobile phone more, the girls or the boys?

4 The mean weight of the 15 men in a slimmers' club is 138.5 kg.
The mean weight of the 13 women in the slimmers' club is 113.3 kg.
(a) Work out the total weight of the men in the club.
(b) Work out the total weight of the women in the club.
(c) Work out the total weight of all the members of the club.
(d) Work out the mean weight of all the members of the club.

Mr and Mrs Jones join the club.
Mr Jones weighs 118.1 kg and Mrs Jones weighs 87.5 kg.
(e) Work out the mean weight of club members now.

5 The shoe sizes of students in a class are shown in this table.

Shoe size	4	5	6	7	8	9
Number of students	4	3	9	8	5	3

(a) How many students are in the class?
(b) Work out the mean shoe size.

6 During the last six years the amounts of petrol used by my car were 1476, 1789, 1689, 2079, 1984 and 1843 litres.
(a) Work out the mean amount of petrol used each year.
(b) How much petrol would I need to use this year in order to reduce my average to 1775 litres?

16.2 The median

1 Write each list in rank order and find the median.
   (a) 10, 6, 14, 14, 11, 7, 8, 4, 17
   (b) 3, 12, 7, 18, 9, 6
   (c) 22, 17, 15, 3, 4, 18, 20, 14, 30, 14, 16, 20, 30, 9
   (d) 216, 432, 319, 325, 285, 371

2 These data show the marks scored by students taking a Maths test.

16	13	27	14	12	25	28	10	24	18	23	12	17	26
11	14	19	14	23	24	16	11	29	21	12	16	19	22

Find the median mark scored in the test.

3 These data show the number of hours spent on homework last week.

Girls: 16 3 7 6 2 5 8 10 4 8 3 5 9 1 9 6 8 7
Boys: 1 14 5 4 3 2 6 11 2 3 4 8

   (a) Work out the median number of hours spent on homework last week for (i) the girls (ii) the boys (iii) the whole class.
   (b) Work out the mean number of hours spent on homework last week for (i) the girls (ii) the boys (iii) the whole class.
   (c) Which average do you think describes the data better? Give a reason for your answer.

4 The salaries, in euro, of 12 workers in a small factory are:

8500	11 000	10 500	10 500	15 000	24 000
10 500	12 000	10 500	8500	11 000	84 000

   (a) Find the median salary.
   (b) Find the mean salary.
   (c) Which average describes the data better? Give a reason for your answer.

5 Davood and Hannah are keen darts players. They record their scores from their last two games. Here are the results.

Davood: 65 70 26 41 39 100 81 57 120 26 30 41
Hannah: 36 180 140 35 41 36 35 25 42 180 15 26 120 15 10

   (a) Find the median score for Davood and Hannah.
   (b) Find the mean score for both of them.
   (c) One of the two players is to be picked to play a match for the local team.
      (i) Give a reason why Davood should be picked.
      (ii) Give a reason why Hannah should be picked.

6 A school secretary recorded the number of letters being posted each day during half a term. Here are the results.

Number of letters	18	19	20	21	22	23	24	25	26	27
Number of days	2	0	5	11	3	4	6	4	3	2

   (a) Find the median number of letters.
   (b) Find the mean number of letters.

16.3 The mode

1 Delyth does a survey of the favourite vegetables of the students in her class. Here are her results.

peas	potatoes	peas	peas	beans	peas	potatoes	peas
onions	potatoes	carrots	potatoes	beans	peas	onions	onions
potatoes	peas	onions	carrots	carrots	beans	potatoes	peas
carrots	cabbage	peas	carrots	potatoes	carrots	potatoes	onions

(a) Draw up a frequency table.
(b) What is the mode?
(c) Why is there no mean or median?

2 Sixteen students watch a cricket match. Their ages are:

11   13   12   16   16   13   12   11   16   11   15   13   12   11   11   11

(a) Find the mode, median and mean for these ages.
(b) Find the range of these data.

3 In a fishing competition the number of fish caught by each angler is recorded. Here are the results.

2   5   0   8   3   4   2   6   2   0   1   4   2
3   2   5   9   7   4   5   5   4   6   6   15

(a) Draw up a frequency table and draw a bar chart to illustrate these data.
(b) Find the mode, median and mean for these data.
(c) Which average best describes the data? Explain your answer.

4 Kathy grows ten tomato plants and uses Brand A fertiliser. She grows another ten plants and uses Brand B fertiliser. She records the number of tomatoes picked from each plant. Here are her results.

Brand A: 21   30   26   21   25   24   28   27   23   26
Brand B: 36   18   14   28   19   35   25   22   27   25

(a) Find the mean number of tomatoes for each brand of fertiliser.
(b) Find the range for each group of tomatoes.
(c) Describe the differences between the two sets of results.

5 Two groups of students play a computer game that tests multiplication tables. These are their scores.

Group A: 186   157   180   158   173   204   192   176   192   170
Group B: 163   152   149   164   163   174   155   196   184   173

(a) Find the mean for each group.
(b) Find the range for each group.
(c) Which group do you think has done better? Explain your answer.

17 Formulae

17.1 Formulae

1 The cost of buying loaves of bread in a supermarket is given by the formula:

Cost (in cent) = 40 × Number of loaves

(a) Find the cost of buying
(i) 1   (ii) 2   (iii) 3   (iv) 4   (v) 5
loaves of bread.
(b) How many loaves of bread can you buy for €4?
(c) The supermarket reduces the price to 35cent per loaf.
Write down the new formula for the cost.

2 Mr Wilson calculates the approximate cost per week of using his car with this formula:

Multiply the number of litres of fuel bought by 4 and then add 10.
The answer is the cost in euro.

(a) Find the cost when he buys
(i) 5   (ii) 7   (iii) 8
litres of fuel.
(b) How much fuel did he buy in a week when he calculated the cost as €34?
(c) Write the formula using €c for the cost and f for the number of litres of fuel.

3 A painter works out the quantity of paint he needs using this formula:

Divide the area to be painted (in m²) by 12 and then add $\frac{1}{2}$.
The answer is the amount of paint in litres.

(a) Calculate the amount of paint he needs for the following areas:
(i) 24 m²   (ii) 60 m²   (iii) 78 m².
(b) What area can he paint with $8\frac{1}{2}$ litres of paint?
(c) Write the formula using A for the area to be painted (in m²) and p for the number of litres of paint.

4 Chocolate Heaven sells boxes of four types of assorted chocolates.
The numbers of each type in a box are:

s strawberry creams
n chocolate nuts
t toffees
f chocolate fudge

The formula for the total cost in cent of each box is

Cost = $6 \times s + 10 \times n + 8 \times t + 9 \times f$

What is the cost of each type of chocolate?

Complete the gaps in the table opposite.

Assortment	s	n	t	f	Total cost
Small	2	2	2	2	?
Medium	?	3	3	4	€1.02
Large	3	3	?	4	€1.16
Super	4	5	5	?	€1.68

17.2 Working with formulae

For this exercise you must remember that it is common to leave out the × signs when you use letters. For example 2a stands for 2 × a.

1 Find the value of $a + b + c$ when:
 (a) $a = 3$, $b = 5$ and $c = 6$
 (b) $a = 10$, $b = 3$ and $c = 12$
 (c) $a = 21$, $b = 13$ and $c = 23$
 (d) $a = 2$, $b = 62$ and $c = 43$
 (e) $a = 17$, $b = 24$ and $c = -12$
 (f) $a = 13$, $b = 26$ and $c = -39$.

2 Find the value of $p + 2q - r$ when:
 (a) $p = 6$, $q = 3$ and $r = 7$
 (b) $p = 12$, $q = 10$ and $r = 5$
 (c) $p = 8$, $q = 9$ and $r = 0$
 (d) $p = 21$, $q = 34$ and $r = 12$
 (e) $p = 30$, $q = 20$ and $r = 70$
 (f) $p = 31$, $q = 45$ and $r = 31$.

3 Use the formula $C = 5d + 6$ to find the value of C when:
 (a) $d = 2$ **(b)** $d = 5$ **(c)** $d = 10$.
 The formula refers to the notice about hiring a wheelbarrow.
 (d) What does d stand for?
 (e) What does C stand for?
 (f) Jake spends €26 on hiring a wheelbarrow. For how many days does he hire it?

Wheelbarrow for Hire.
€5 per day

plus
€6 handling charge.

4 Julia uses this formula for converting degrees Centigrade (C) to degrees Fahrenheit (F):

$$F = 1.8C + 32$$

Calculate the temperature in Fahrenheit when the temperature in Centigrade is:
 (a) 15°   **(b)** 20°   **(c)** 37°   **(d)** 25°.
 What is the temperature in Centigrade when the temperature in Fahrenheit is:
 (e) 32°   **(f)** 50°?

5 **Investigation**

Mr Williams is planning a small car and lorry park.

> The parking space needed for a car is $12\,\text{m}^2$.
> The parking space needed for a lorry is $30\,\text{m}^2$.

He uses the formula $A = 12c + 30l$ to work out the total area needed.
 (a) What do A, c and l stand for?
 (b) Find three possible combinations that include both car and lorry spaces and use $240\,\text{m}^2$ altogether.
 (c) Mr Williams plans to charge €5 per day for a car and €12 per day for a lorry. Write a formula for the total daily takings (T).
  If all the available spaces are used which of your answers to **(b)** would give the highest taking?

18 Negative numbers

18.1 Addition of negative numbers

1 These were the noon temperatures one day in January.

Belgrade	$-4\,°C$
Paris	$5\,°C$
Sydney	$28\,°C$
Houston	$11\,°C$
Moscow	$-10\,°C$

(a) Which place had the lowest temperature?
(b) Draw a thermometer marked from $-10°$ to $30°$.
    Mark each of these cities by the correct temperature.
(c) Add these cities to your diagram:

Toronto	$-1\,°C$
Honolulu	$23\,°C.$

2 Draw a number line from -10 to 10.
Use it to help you find the number which is:
(a) 3 less than 7
(b) 2 more than -1
(c) 3 less than -2
(d) 5 less than 3
(e) 4 more than -9
(f) 7 more than -10.

3 Work out the following.
(a) $(-2) + (+5)$
(b) $(-2) + (-5)$
(c) $(+2) + (-5)$
(d) $(+3) + (-7)$
(e) $(-4) + (+6)$
(f) $(-4) + (-6)$
(g) $(-6) + (-9)$
(h) $(+7) + (-27)$
(i) $(-3) + (+50)$

4 Work out the following.
(a) $-6 + 8$
(b) $5 - 7$
(c) $-5 + 7$
(d) $-4 + 16$
(e) $-6 + 12$
(f) $-3 + 5 + 6$
(g) $4 - 6 + 10$
(h) $-3 + 9 - 15 + 2$
(i) $-5 + (-6) + 7 + 8$

5 Use your calculator to answer this question. This checks that you can use your calculator with negative numbers.
Work out the following.
(a) $6 + (-2)$
(b) $-7 + 6$
(c) $-3 + (-7)$
(d) $-5 + 22$
(e) $-200 + 190$
(f) $-18 + (-32)$

6 Find the missing number.
(a) $-2 + \square = -6$
(b) $5 + \square = -3$
(c) $6 + \square = 10$
(d) $10 + \square = -15$
(e) $\square + 6 = -1$
(f) $\square + (-3) = -30$

18.2 Subtraction of negative numbers

1 These were the noon temperatures one day in January.

Belgrade $-4\,^{\circ}$C
Paris $5\,^{\circ}$C
Sydney $28\,^{\circ}$C
Houston $11\,^{\circ}$C
Moscow $-10\,^{\circ}$C

(a) How much warmer was Sydney than Paris?
(b) How much warmer was Paris than Belgrade?
(c) What was the difference between the temperatures of Moscow and Belgrade?
(d) What was the temperature difference between the coldest and warmest cities?

2 Put these numbers in order of size, smallest first.
(a) $6, -2, 9, -5, 1, -3$
(b) $-3, 16, 94, -100, 4$
(c) $22, -28, -7, 41, -50$

3 Put these numbers in order of size, largest first.
(a) $1, -5, 3, -6, 7, -2$
(b) $2, 17, -8, 26, -13, -30$
(c) $14, 140, -15, -150, -1.4, 1.5$

4 Complete these calculations.
(a) $6 - (-4) = 6 + 4 = \Box$
(b) $5 - (-7) = 5\ \Box\ 7 = \Box$
(c) $4 - \Box = 4 + 3 = 7$

5 Work out (evaluate) the following.
(a) $5 - 2$ (b) $6 - 10$ (c) $3 - 9$
(d) $-2 - 1$ (e) $-5 - 3$ (f) $6 - (-2)$
(g) $4 - (-3)$ (h) $-2 - (-1)$ (i) $-9 - (-3)$

6 Evaluate the following.
(a) $7 - 4$ (b) $6 - 8$ (c) $2 - 5$
(d) $-2 - 3$ (e) $-4 - 3$ (f) $6 - (-3)$
(g) $5 - (-4)$ (h) $-5 - (-1)$ (i) $-4 - (-3)$

7 Evaluate the following.
(a) $7.2 - 4.5$ (b) $6.3 - 8.0$
(c) $2.0 - 5.4$ (d) $-2.5 - 3.2$
(e) $-4.5 - 3.3$ (f) $6.6 - (-3.4)$

8 Azdil has these cards.

$\boxed{6}$ $\boxed{2}$ $\boxed{-4}$ $\boxed{8}$ $\boxed{5}$ $\boxed{-7}$ $\boxed{-10}$

(a) Which two cards give the largest answer in this calculation?
     $\Box + \Box =$
(b) Which two cards give the smallest answer in this calculation?
     $\Box + \Box =$
(c) Which two cards give the largest answer in this calculation?
     $\Box - \Box =$
(d) Which two cards give the smallest answer in this calculation?
     $\Box - \Box =$

18.3 Multiplying and dividing negative numbers

1 Work these out without your calculator.
- **(a)** $(+2) \times (-3)$
- **(b)** $(+5) \times (-2)$
- **(c)** $(-6) \times (2)$
- **(d)** $2 \times (-7)$
- **(e)** $(-5) \times (-3)$
- **(f)** $(-6) \times (-4)$

Then use your calculator to check your answers before continuing with question 2.

2 Evaluate the following.
- **(a)** $(-3)^2$
- **(b)** $(-1)^2$
- **(c)** 0×-5
- **(d)** $(-7) \times (-8)$
- **(e)** $12 \times (-3)$
- **(f)** $(-5) \times 0$
- **(g)** $(-4) \times 5$
- **(h)** $(-9) \times (-4)$
- **(i)** $(-6)^2$
- **(j)** $(+5) \times (+3)$
- **(k)** $11 \times (-2)$
- **(l)** $(-4) \times 8$
- **(m)** $0 \times (-7)$
- **(n)** $(-5)^2$
- **(o)** $(-8) \times (-20)$

3
$$\boxed{(-5) \times (-3) = (+15)} \qquad \boxed{(+4) \times (-12) = (-48)}$$

Use these rules to work out the following.
- **(a)** $(+15) \div (-3)$
- **(b)** $(+15) \div (-5)$
- **(c)** $(-48) \div (+4)$
- **(d)** $(-48) \div (-12)$

4 Evaluate the following.
- **(a)** $6 \div (-3)$
- **(b)** $20 \div (-5)$
- **(c)** $(-40) \div (+4)$
- **(d)** $(-6) \div (-2)$
- **(e)** $(-12) \div (-4)$
- **(f)** $18 \div (-2)$
- **(g)** $(-16) \div 2$
- **(h)** $(-28) \div (-4)$
- **(i)** $(-8) \div (-8)$
- **(j)** $14 \div (-7)$
- **(k)** $(-6) \div 6$
- **(l)** $(-60) \div 4$
- **(m)** $45 \div (-5)$
- **(n)** $(-100) \div (-20)$
- **(o)** $(-62) \div 62$

5 Evaluate the following.
- **(a)** $(-12) \times (-3)$
- **(b)** $(-12) \div (-3)$
- **(c)** $(-5) \times (+6) \times (-2)$
- **(d)** $(-9)^2$
- **(e)** $8 \times (-3) \times 2$
- **(f)** $27 \div (-3)$
- **(g)** $(-40) \div 8$
- **(h)** $(-6) \times 7 \times (-2)$
- **(i)** $(-2) \times (-5) \times (-3)$
- **(j)** $44 \div (-11)$
- **(k)** $(-3)^3$
- **(l)** $52 \div (-4)$
- **(m)** $(-72) \div (-9)$
- **(n)** $7 \times (-5) \times (-4)$
- **(o)** $(-4)^3$

6

$\boxed{6}$ $\boxed{2}$ $\boxed{-4}$ $\boxed{8}$ $\boxed{-12}$ $\boxed{-10}$ $\boxed{5}$

Which two of these numbers give:
- **(a)** the largest answer when multiplied together
- **(b)** the smallest answer when multiplied together
- **(c)** the largest answer when one is divided by the other
- **(d)** the smallest answer when one is divided by the other?

19 Equations

19.1 Balancing equations

1 Write down the equation for each of the following sets of scales.
Work out how much each package weighs. (All the unmarked packages in one diagram are the same weight.)
Draw the scales at each stage in your working.

(a) 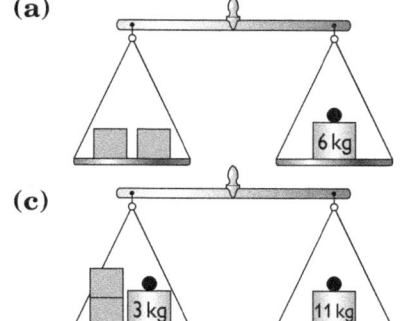 **(b)**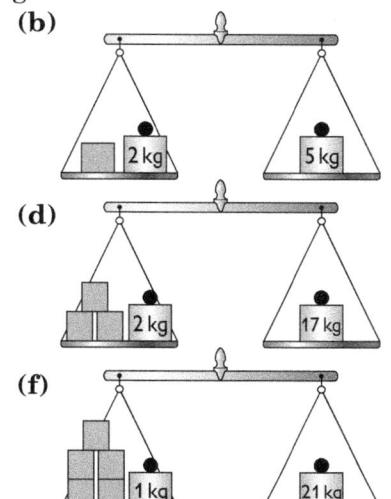

(c)

(d)

(e)

(f)

2 Draw the scales for each of these equations.
Work out how much each package weighs.
Draw the scales at each stage in your working.
(a) The weight of 3 packets balances 18 kg.
(b) The weight of 2 packets and 5 kg balances 17 kg.
(c) The weight of 4 packets and 3 kg balances 27 kg.
(d) The weight of 5 packets and 5 kg balances 25 kg.

19.2 Solving equations

1 Solve these equations by subtracting from both sides.
(a) $x + 5 = 19$ **(b)** $a + 3 = 8$ **(c)** $4 + y = 11$

2 Solve these equations by adding to both sides.
(a) $p - 2 = 1$ **(b)** $r - 4 = 3$ **(c)** $t - 11 = 19$

3 Solve these equations.
(a) $x + 6 = 9$ **(b)** $y + 4 = 6$ **(c)** $s - 5 = 10$ **(d)** $t + 12 = 16$ **(e)** $p - 3 = 5$

4 Solve these equations.
(a) $3a = 12$ **(b)** $2x = 10$ **(c)** $5y = 30$

5 Solve these equations.
(a) $4n + 3 = 11$ **(b)** $2x - 7 = 11$ **(c)** $3y - 5 = 16$ **(d)** $4p + 2 = 22$
(e) $10s - 11 = 9$ **(f)** $8t + 5 = 69$ **(g)** $7z + 11 = 18$

6 Three identical packages and a trolley weigh 52 kg.
The trolley weighs 34 kg. How much does each package weigh?

7 **Investigation**

'I think of a number, double it and subtract 5. I finish with the same number as I started with. What is my number?'
Make up some more puzzles like this.

19.3 Solving problems with equations

1 Three children's tickets and one adult ticket to a cinema cost €14. The adult ticket costs €5. A child's ticket costs €c. Form an equation in c and solve it.

2 A school buys 12 scientific calculators and three graphical calculators for €168. The graphical calculators cost €32 each. A scientific calculator costs €s. Form an equation in s and solve it.

3 Sarah is 4 years older than Claire. Their combined ages are 36 years. Sarah's age is s years. Form an equation in s and solve it.

4 Trevor walks a circuit of c km once each day from Monday to Friday. At the weekend he walks to town and back once, a distance of 12 km. In that week he walks a total distance of 47 km. Form an equation in c and solve it.

5 A small ferry has 12 cars and 23 passengers aboard. Each passenger pays €1. The total collected is €59. The cost for a car is €w. Form an equation in w and solve it.

6 In a game there are 2 points for a draw and 1 point for a loss. One team's results are

W	D	L
7	2	3

They have got 42 points. Form an equation and use it to find out how many points there are for a win.

20 Solid shapes

20.1 Nets

1 Draw the net of a cube.

2 A die is in the shape of a cube.
Opposite faces of a die add to 7.
(a) Look at these nets of dice. Work out what numbers go on the lettered faces.

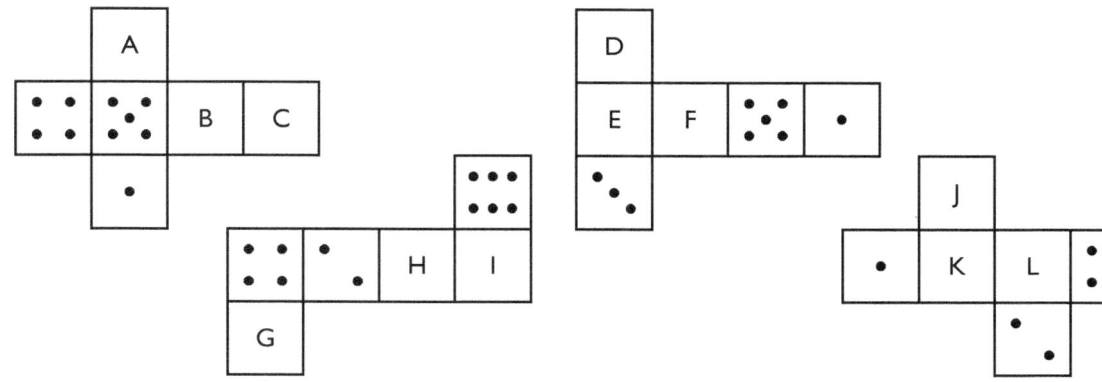

(b) Which of the nets below are numbered correctly?
For a net incorrectly numbered, say which numbers need exchanging to make it right.

(i)
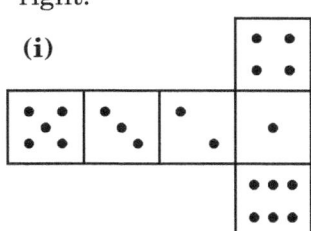

(ii)
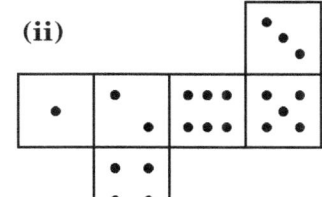

(iii)
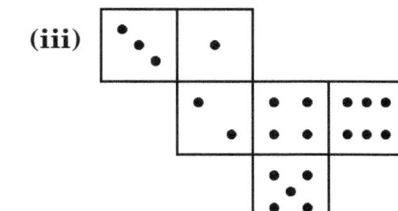

(iv)
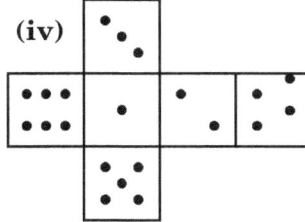

3 Look at this cuboid.
(a) Write down all the edges that are parallel to AB.
(b) Write down all the faces that are perpendicular to face ABCD.
(c) Which three faces meet at vertex G?

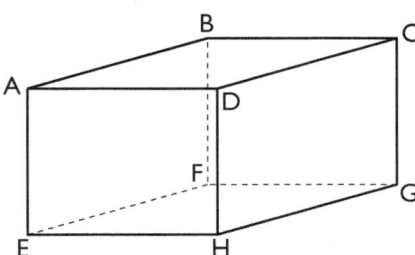

4 *You will need some centimetre triangular dotty paper for this question.*
This cuboid is drawn on centimetre triangular dotty paper.
(a) **(i)** Write down the length, width and height of the cuboid.
　　(ii) Draw an accurate net of the cuboid.
(b) Draw, on centimetre triangular dotty paper, the cuboid with dimensions 4 cm by 3 cm by 2 cm.

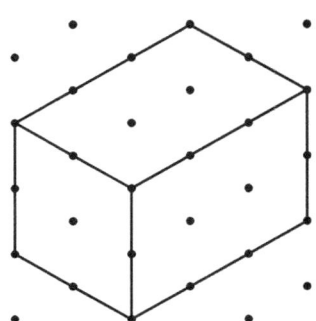

20.2 Prisms

1 Which of these models are prisms?

(a) 　**(b)** 　**(c)** 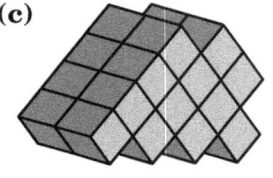　**(d)**

2 Which of these objects are prisms?

(a) 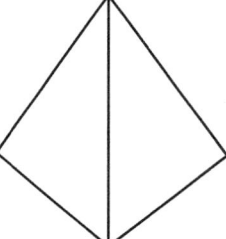　**(b)** 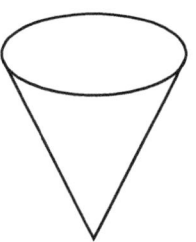　**(c)**

(d) 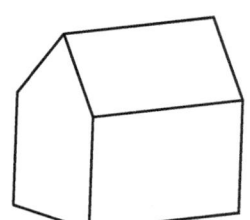　**(e)** 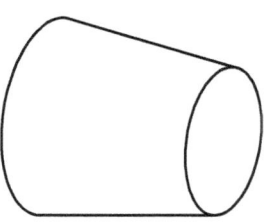　**(f)**

(g) 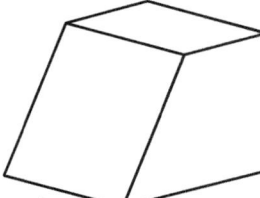　**(h)**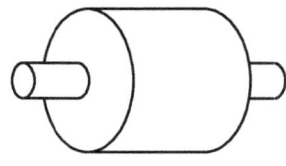

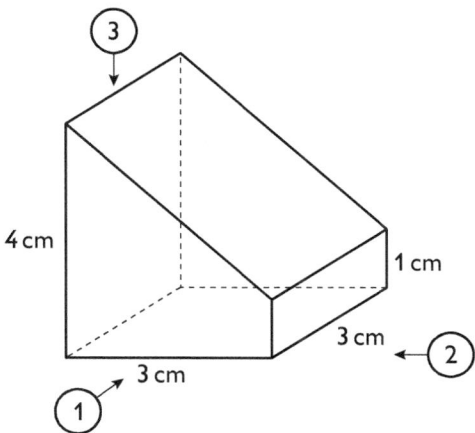

(a) Draw an accurate view of this prism from each of the directions shown.
(b) Use your diagram of view 1 to find the length of the sloping face of the prism.
Hence draw an accurate net of the prism.

4 This object is also a prism. A prism can be defined as:
'A solid with its ends congruent and parallel.'
Draw a sketch of another prism where the sides are not at
right-angles to the ends.

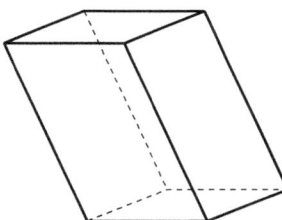

5 Investigation

List some everyday objects that are prisms.
Find pictures in magazines of objects that are prisms.

20.3 Further solid shapes

1 Name seven different types of solid shapes.

2 There are five platonic solids.
Name them all.

3 Draw accurately a net of this square-based pyramid.
Use your net to find the unknown length of the sides of the
triangular faces.

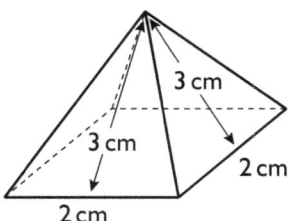

4 Which of these is the net of a pyramid?
In the cases where the net will not form a pyramid, explain why not.

(a) 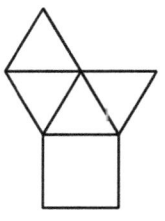   **(b)** 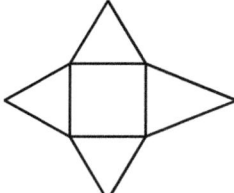   **(c)**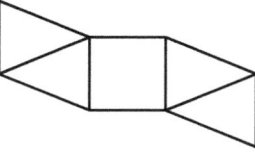

(d) 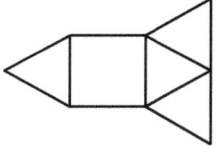   **(e)**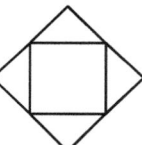

5 This is the net of an octahedral-shaped die.
The 'opposite' numbers add to 9.
Make a sketch of the net and put on the
numbers.
(You may find this easier if you make the
model.)

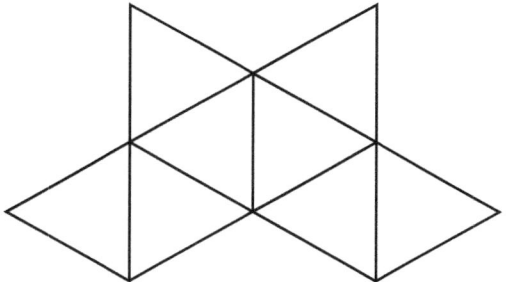

6 On a piece of paper, draw a circle of radius 5 cm.
Cut out a sector of the circle.
This gives the net of a cone.
Stick the straight edges together with tape and make the cone.

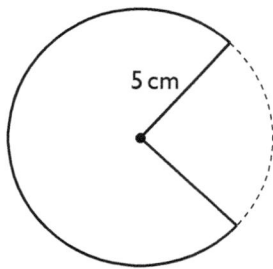

21 Graphs

21.1 The equation of a line

For this exercise you will need squared paper or graph paper.

1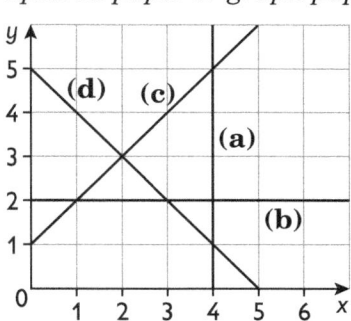

Write the co-ordinates of five points on each line in the graph above.
For each line, look at the patterns in the numbers. Then find the equation of the line.

2 Write the co-ordinates of four points on each of these lines. Then draw the lines.

(a) $x = 5$ **(b)** $x + y = 6$

(c) $y = -2$ **(d)** $y = x - 3$

(e) $y = 5 - x$ **(f)** $x + y = 0$

3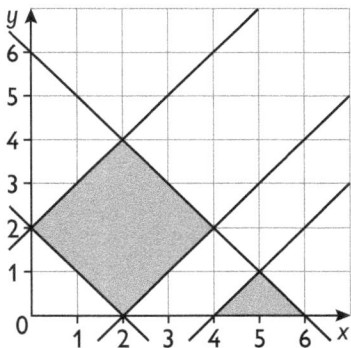

(a) Find the equations of the four lines which make the shaded square.

(b) Find the equations of the three lines which make the shaded triangle.

4 For each set of points:

(i) look at the patterns in the numbers

(ii) write three more points on the same line

(iii) draw the line

(iv) check that your points are correct and label the line with its equation.

(a)	**(b)**	**(c)**
$(2, 6)$	$(5, 5)$	$(7, 2)$
$(3, 7)$	$(4, 4)$	$(6, 1)$
$(4, 8)$	$(3, 3)$	$(5, 0)$
$(5, 9)$	$(2, 2)$	$(4, -1)$

21.2 Drawing lines

For this exercise you will need squared paper or graph paper.

1 (a) Copy and complete this table.

x	-4	-3	-2	-1	0	1	2	3	4
$y = x + 3$	-1			2				6	

Write down the largest and smallest values of both x and y.
Then draw axes and the graph.

(b) Use your graph to find the value of y when $x = 1.5$ and the value of x when $y = 1.6$.

2 (a) Copy and complete this table.

x	-4	-3	-2	-1	0	1	2	3	4
$y = x - 4$		-7			-4			-1	

Then draw the graph.

(b) Compare your graph with question 1. What do you notice?

3 Draw axes with x from -4 to 4 and y from -12 to 12.
You are going to draw the six lines **(a)** to **(f)** below on the same set of axes.
Make a table of values for each one.
Then draw the line and label it.

(a) $y = x$ (b) $y = 2x$

(c) $y = 3x$ (d) $y = -x$

(e) $y = -2x$ (f) $y = -3x$

What do you notice about these lines?

4 (a) Copy and complete this table.

x	-4	-3	-2	-1	0	1	2	3	4
$3x$	-12			-3			6		
$+1$	$+1$			$+1$			$+1$		
$y = 3x + 1$	-11			-2			7		

(b) Make similar tables for $y = 3x + 4$ and $y = 3x - 3$.

(c) Draw the graphs for all three tables on the same set of axes.

(d) Add the graph of $y = 3x + 2$ to the same axes. How can you do this without making another table of values?

21.3 Interpreting graphs

For this exercise you will need graph paper.

1 This graph is used for converting between pounds and euro for the exchange rate on one day.
Use the graph to find:
  (a) how many euro you get for
      (i) £20
      (ii) £30
  (b) how many pounds you get for
      (i) €50
      (ii) €70.

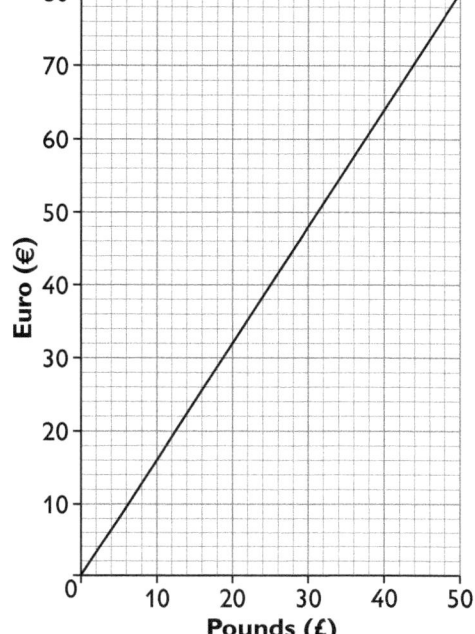

2 This graph shows the second-hand value of a car which was bought when new.

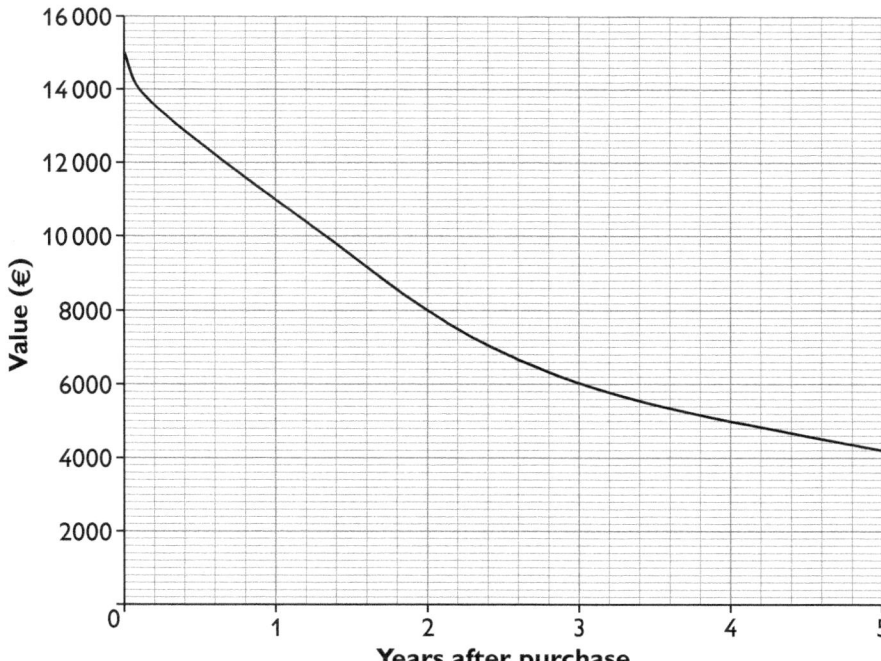

  (a) What was the value when new?
  (b) What does the steep section of the graph just after purchase show?
  (c) By how much did the value fall in the first year?
  (d) What was the value after $2\frac{1}{2}$ years?
  (e) When was the value €5000?

3 Kali cycled from home to Sharon's house.
The graph shows her journey.

(a) She stopped on the way. How long did she stop for?

(b) How far from Kali does Sharon live?

(c) What was Kali's speed on the fastest part of her journey to Sharon's house?

(d) The last section of the graph slopes downwards. What does this show?

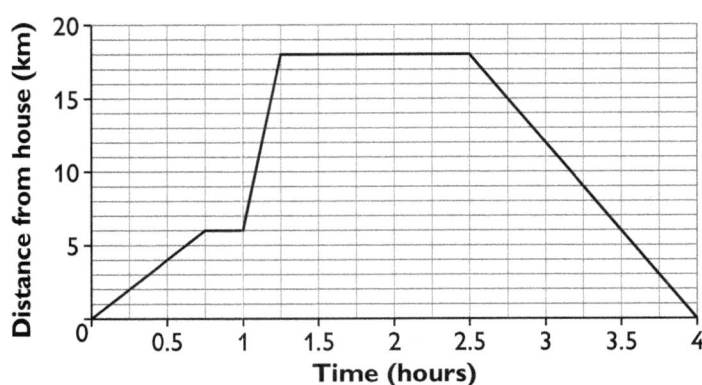

4 This graph shows the noon temperature in Ashton and Beeby for one week.

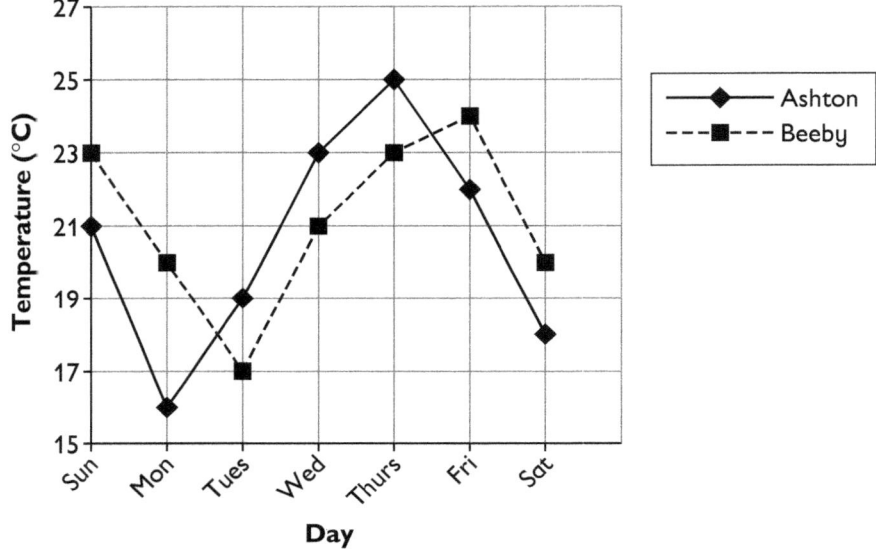

(a) Which town had the highest temperature?

(b) For how many days was Ashton warmer than Beeby?

(c) What was the difference between their temperatures on Thursday?

(d) Find the difference between Ashton's temperature on Tuesday and Wednesday.

5 Anna started to run water for a bath. After two minutes, there were 25 litres of water in the bath. She turned the taps on faster, and the bath filled with another 20 litres in the next minute. She took five minutes to have her bath, then let the water out, which took a minute. Draw a graph to show the volume of water in the bath during this time.

6 Investigation

Make up a 'time story' in the style of question 5. On a separate piece of paper draw a graph for your story.

Swap your story with a friend and draw each other's graphs. Check that the two graphs agree with each other in each case.

22 Measuring

22.1 Perimeters, area and volume

1 Work out the area and perimeter of this rectangle.

2 Copy and complete the table below for each of the rectangles.

	Length	Width	Area	Perimeter
(a)	4 cm	5 cm		
(b)	8 cm		48 cm²	
(c)	10 cm			32 cm

3 The area of a square is 36 cm². What is the perimeter of the square?

4 The floor of a school hall is a rectangle.
The length of the hall is 10 m more than the width.
The perimeter of the hall is 140 m.
What is the area of the floor?

5 The area of a rectangular driveway is 120 m².
The perimeter of the driveway is 68 m.
Find the length and width of the driveway.

6 Copy and complete the table below for each of the cuboids.

	Length	Width	Height	Volume
(a)	9 cm	5 cm	2 cm	
(b)	3 m	6 m	1.5 m	
(c)		2 cm	3 cm	48 cm³
(d)	0.4 m		6 m	1.2 m³
(e)	8 cm	6 cm		144 cm³

7 What length of wire would be needed to make a skeleton cuboid 25 cm long, 18 cm wide and 7 cm high?

8 A metal water tank is in the shape of a hollow cuboid with no top.
The water tank measures 5 m by 3 m by 2 m.
What area of metal is needed to make the tank?

22.2 More areas

1 Calculate the area of each of these triangles.

(a)

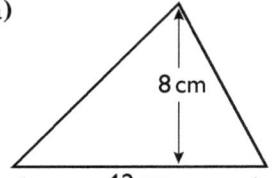

(b)

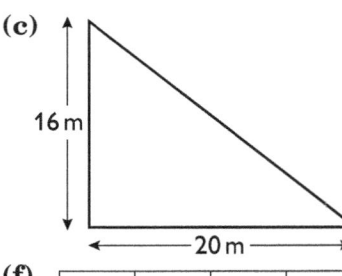

(c)

(d)

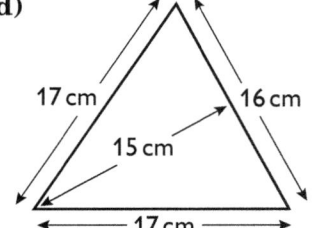

(e)

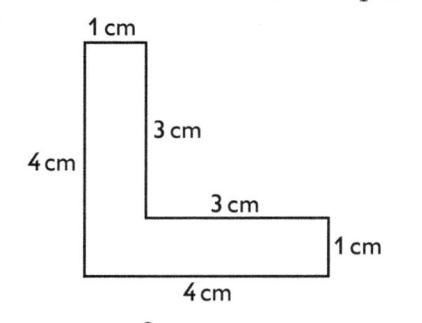

(f)

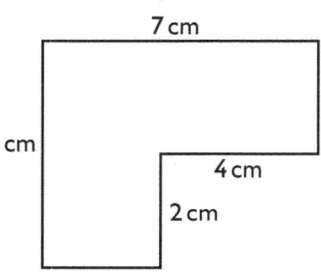

2 Calculate the area of these shapes by splitting them into rectangles and triangles.

(a)

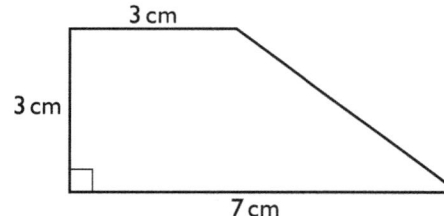

(b)

(c)

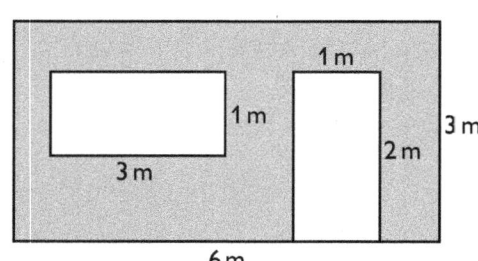

3 Andy wants to paint the side wall of his garage.
Work out the area to be painted.

4 Work out an estimate of the area of each of these shapes.

(a) **(b)**

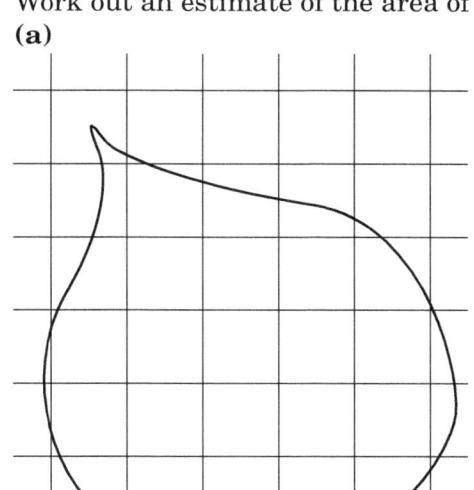 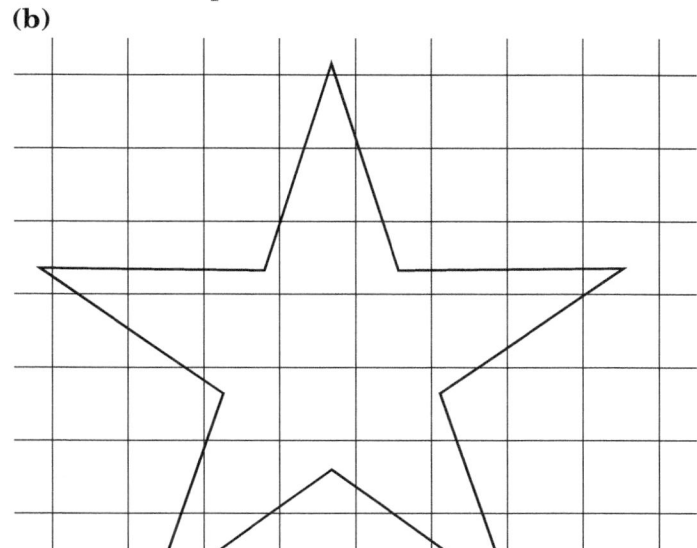

22.3 Volume of prisms

1 Calculate the volume of these prisms.

(a) **(b)**

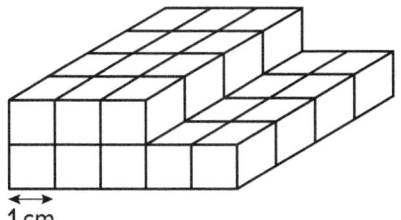 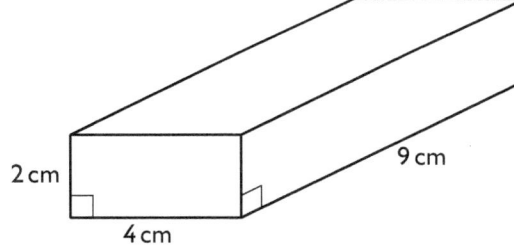

(c) **(d)**

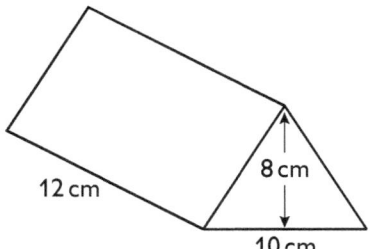 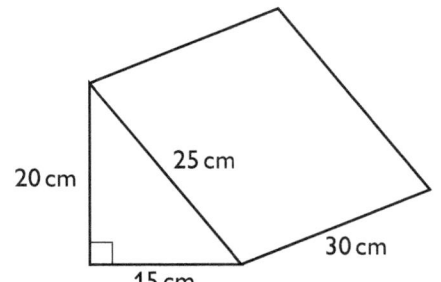

2   Calculate the volume of these prisms.

(a)

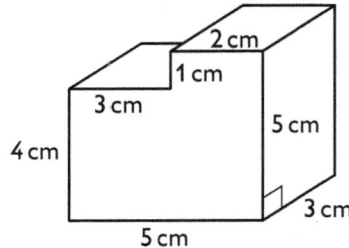

(b)

(c)

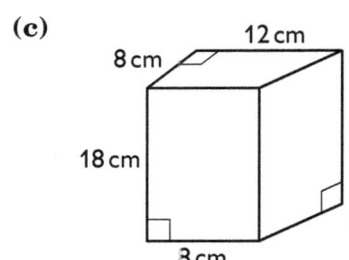

3   A pond has a surface area of 200 m².
Ice, 10 cm thick, forms over the surface of the pond.
(a)   What is the volume of ice on the pond?
(b)   1 m³ of ice weighs about 900 kg.
Work out the weight of the ice on the pond.

22.4 More volumes

1 A greenhouse is in the shape of a triangular prism.
   (a) Find the volume of the greenhouse.
   (b) Find the total area of the four glass walls.

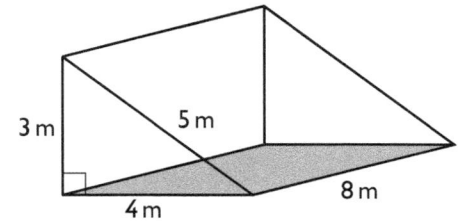

2 ABCD is the outline of a field.
   (a) Work out the area of the field.
   (b) Soil to a depth of $\frac{1}{2}$ m is scraped off the surface of the field.
       What volume of soil is removed?

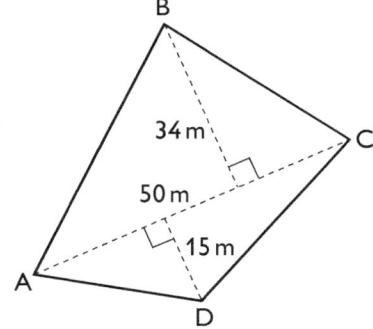

3 A water tank measuring 6 m by 3 m by 4 m is full of water.
   (a) Find the volume of water in the tank.
   (b) 54 m^3 of water is taken from the tank.
   What is the depth of water in the tank now?

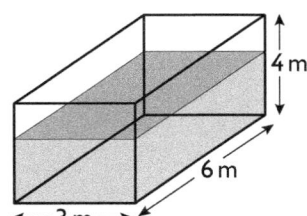

4 The diagram shows the side of a swimming pool.
   The pool is 10 m wide.
   Work out the volume of water needed to fill the pool.

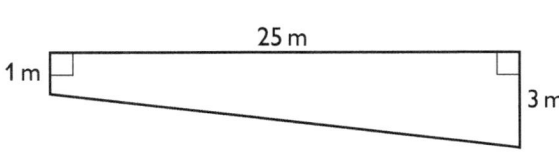

5 The picture shows two cubes.
   Cube A has side of length 3 cm and cube B has side of length 6 cm.
   (a) By how many times is the length of B larger than the length of A?
   (b) By how many times is the surface area of B larger than the surface area of A?
   (c) By how many times is the volume of B larger than the volume of A?
   (d) Write down the connection between your answers to **(a)**, **(b)** and **(c)**.

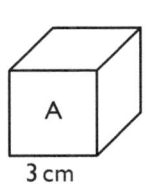
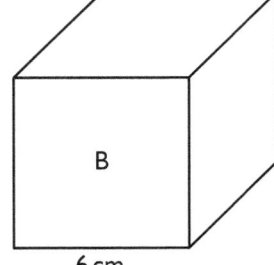

23 Flow diagrams and number patterns

23.1 Decision tree diagrams

1 The flow chart on the right sorts
these types of water transport:

canoe   *car ferry*
cruise liner   *container ship*
rowing boat

Find the missing questions and
names.

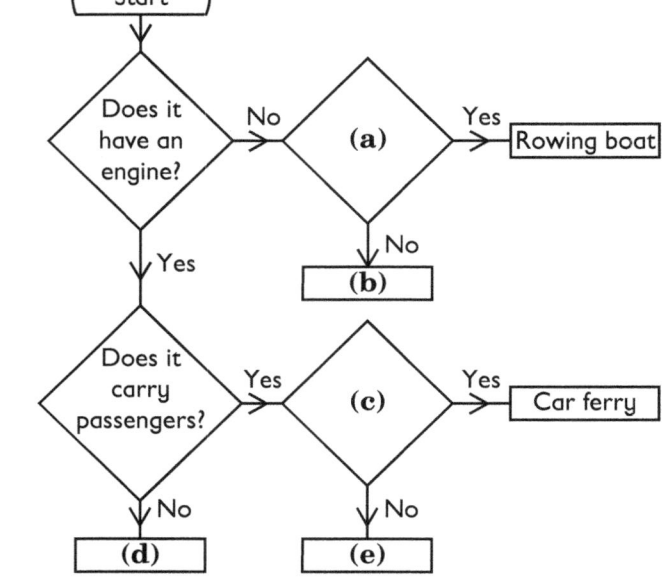

2 This flow chart sorts
shapes with four sides.
The five boxes contain
the names kite,
parallelogram,
rectangle, rhombus and
square. Find which one
goes where.

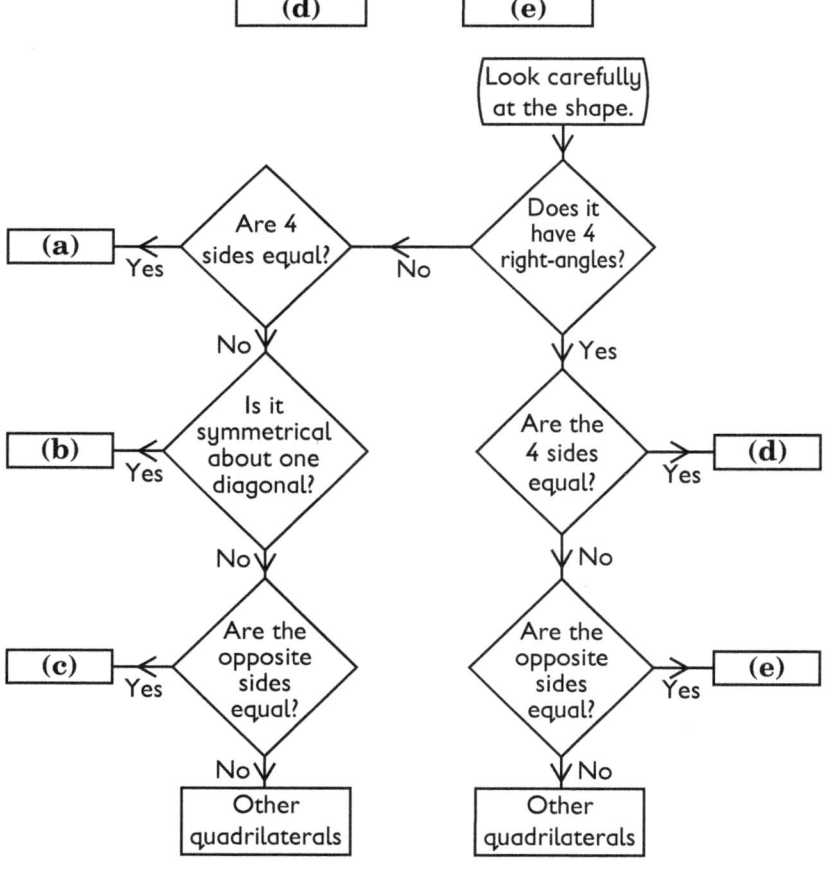

23.2 Flow charts and number patterns

1 Work through this flow chart.

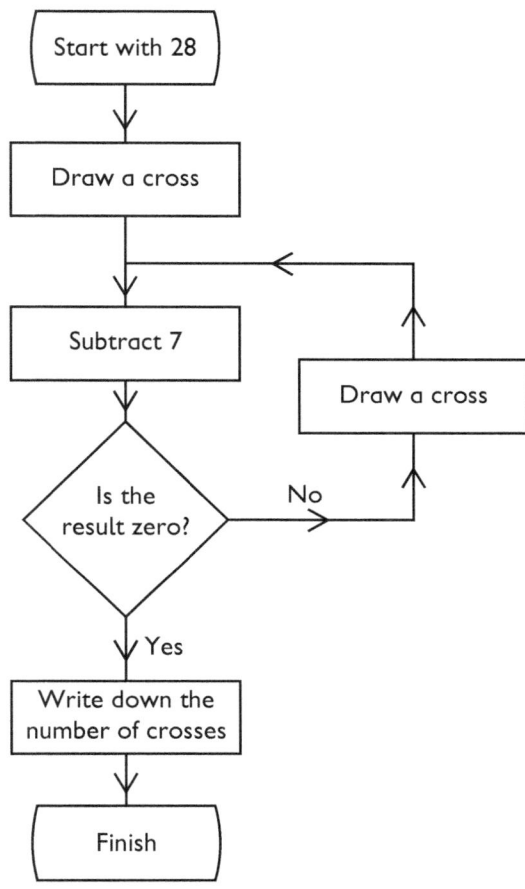

(a) How many crosses did you draw?

(b) What does the flow chart do?

(c) Draw a flow chart for dividing 20 by 5.

(d) What happens if you try to divide 13 by 4?

(e) Change the question box so that the flow chart works when there is a remainder.

2 Design a flow chart to give this number pattern.

1   4   7   10   13.

3 **Investigation**

Design a flow chart which only gives the 'remainders'.

23.3 Working systematically with patterns

1 **(a)** How many different two-digit numbers can you make by choosing two of the digits 1, 2 or 3?

(b) Write down all the different three-digit numbers using each of the digits 1, 2 and 3 just once in each number. How many numbers can you find?

(c) Here is Safiq's working.

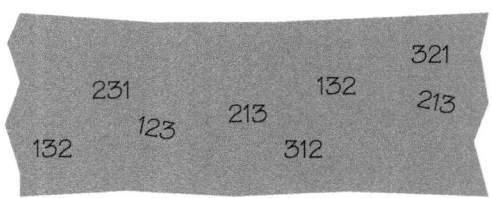

Is his working correct?
How could he have prevented himself making slips?

2 *You may need some 1cent and 2cent coins for this question.*
If you use only 1cent and 2cent coins there are three different ways of making a total of 5cent, as shown.

(a) Find the number of ways of making:

(i) 4cent	**(ii)** 6cent
(iii) 2cent	**(iv)** 8cent
(v) 3cent	**(vi)** 7cent
(vii) 11cent	**(viii)** 12cent
(ix) 9cent	**(x)** 10cent.

(b) Use your results in **(a)** to help you complete this table – without using coins.

Total	13cent	14cent	15cent	16cent	17cent	18cent	19cent
Number of ways							

3 Repeat question 2, but this time use 1cent, 2cent and 5cent coins.

4 *You will find it helpful to use some triangular dotty paper for this question.*
How many different patterned hexagon tiles can you make by using two colours?
Here are three:

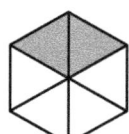   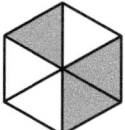   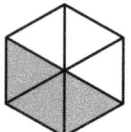

23.4 Growth patterns

You will need some squared paper for this exercise.

For questions 1 and 2:

(a) copy and complete the table
(b) predict the number of black squares in the next two stages
(c) check your predictions.

1

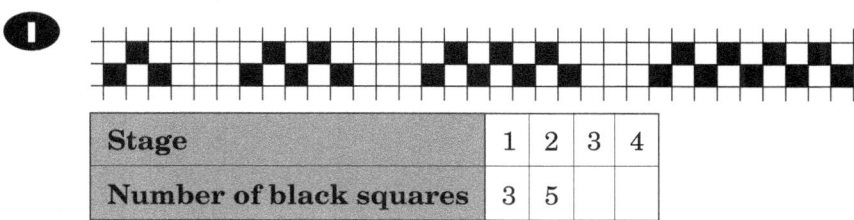

Stage	1	2	3	4
Number of black squares	3	5		

2

Stage	1	2	3	4
Number of black squares	8			

3 These patterns are made out of black and white tiles.

(a) There are nine tiles in the first stage.
How many are there in the second stage?
(b) How many tiles will be needed in the sixth stage?
(c) Look at the stages in the growth pattern.
How many black tiles would you expect there to be in the next two stages?
Check your prediction by drawing the next two stages.
(d) How many white tiles would you expect there to be in the next three stages?
Check your answer yourself.

4 **Investigation**

For each of these growth patterns, find a pattern with shaded squares.
(a) 1   5   9   13   ...   **(b)** 2   6   12   20   ...

23.5 Number patterns – making predictions

1 Here is a pattern of black squares.

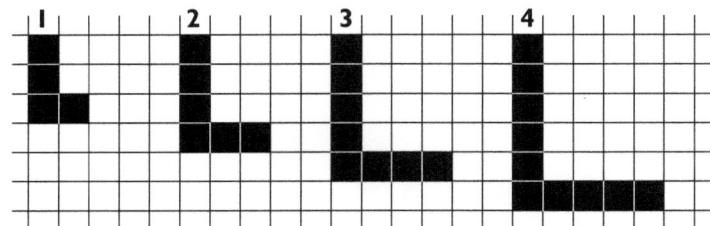

(a) Complete this table for the first four patterns.

Pattern	Number of squares
1	4
2	
3	
4	

(b) How many black squares are there in the
  (i) 5th **(ii)** 6th **(iii)** 50th **(iv)** 60th pattern?

(c) Write down the rule you used to answer **(b)**.

2 Write down a rule connecting the number of grey tiles with the number of black tiles for this growth pattern.

3 For each of these rules draw a pattern of black squares that fits it.
  (a) Number of squares = 2 × pattern number.
  (b) Number of squares = 1 + pattern number.
  (c) Number of squares = pattern number × pattern number.

4 **Investigation**

Investigate the number of dots in this growth pattern.
Allow yourself about ten minutes, and make a note of anything interesting you find.

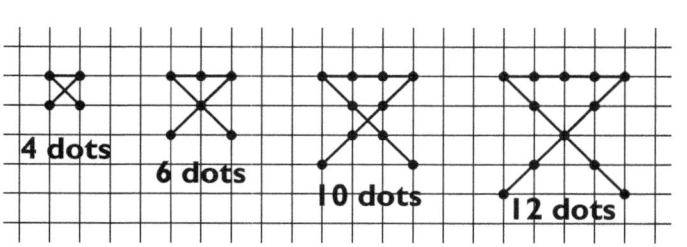

4 dots 6 dots 10 dots 12 dots

24 Accuracy

24.1 Division (dealing with remainders)

1 Mum makes 1000 ml of jam.
She puts it into jars each of which holds 80 ml of jam.
(a) How many jars does she fill?
(b) How much jam is left?

2 Water pipes are 3 m long.
How many lengths of pipe are needed to cover a distance of 71 m?

3 Rebecca's birthday is 113 days after Owen's birthday.
How many full weeks are there between their birthdays?

4 A fence panel, 92 cm wide, is made by joining five boards of equal width.
How wide is one of the boards?

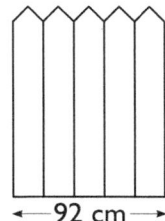

←—92 cm—→

5 Tins of beans are sold in boxes of 11.
A supermarket wants to order 270 tins of beans.
(a) How many boxes must they buy?
(b) How many extra tins will there be?

6 For each of the following, write your answer, giving any remainder as a fraction.
(a) $94 \div 10$ (b) $22 \div 8$ (c) $17 \div 6$
(d) $159 \div 12$ (e) $204 \div 15$

7 (a) €85 is shared equally between Emma and Rebecca.
How much do they each get?
(b) Four identical books cost €21.
How much does one of the books cost?
(c) Five loaves of bread last a family three days.
On average, how much bread is eaten each day?
(d) Jack raises €18.50 for charity.
He divides the amount equally between three charities.
How much does each charity receive?

24.2 Rounding (giving numbers to a sensible degree of accuracy)

1 Re-write the following statements using sensible approximations for the numbers used.
(a) It is 346.7 km from London to Liverpool.
(b) The height of Mount Everest is 8817 m.
(c) The Daily Record sold 341 561 newspapers yesterday.
(d) 19 578 spectators watched the first test match.
(e) The new motorway has cost €68 412 603.

2 Copy and complete the table below.
Round each of the numbers to the accuracy stated.

	Number	To the nearest 10	To the nearest 100	To the nearest 1000
(a)	7359			
(b)	6047			
(c)	1555			
(d)	3456			
(e)	8499			
(f)	76 543			

3 Round 673 843 to the nearest:
(a) 10 **(b)** 100 **(c)** 1000 **(d)** 10 000 **(e)** 100 000 **(f)** 1 000 000.

4 **(a)** A boy's height is given as 160 cm, correct to the nearest 10 cm.
Between what limits does his height lie?
(b) The boy's weight is 90 kg, correct to the nearest 10 kg.
Between what limits does his weight lie?
(c) The boy has €4000 in his bank account, correct to the nearest €100.
Between what limits does this amount of money lie?
(d) The boy's home is 12 000 m from school, correct to the nearest 1000 m.
Between what limits does this distance lie?

24.3 To the nearest whole number or whole unit

1 Round these measurements to the nearest centimetre.
(a) 6.1 cm **(b)** 2.8 cm **(c)** 4.47 cm **(d)** 0.9 cm **(e)** 4.5 cm
(f) 7.7 cm **(g)** 4.09 cm **(h)** 9.6 cm **(i)** 5.45 cm **(j)** 1.297 cm
(k) 63.49 cm **(l)** 139.75 cm **(m)** 0.23 cm **(n)** 1000.06 cm **(o)** 6.277 m

2 **(a)** **(i)** Add together 5.2, 4.4, 9.9, 2.1.
   (ii) Round your answer to the nearest whole number.
(b) **(i)** Round each of the numbers in **(a)(i)** to the nearest whole number.
   (ii) Add these answers together.
(c) Explain why there is a difference between your answers to **(a)(ii)** and **(b)(ii)**.

3 Work out the following money calculations.
Give your answers correct to the nearest cent.
(a) 97cent ÷ 3 **(b)** €1.55 ÷ 6 **(c)** €2840 ÷ 7 **(d)** €3.53 ÷ 2

4 Fifteen people win €2 361 894 on the Lottery.
They share the winnings equally between them.
(a) How much do they each receive, to the nearest euro?
(b) Is there a problem with rounding the amount they receive to the nearest euro?

5 Are 5 cm and 5.0 cm different measurements?
Explain your answer.

6 Copy and complete the table below.
Each quantity has been written to the nearest whole number.
Work out the lowest and highest possible values that it could be.

	Quantity	Highest possible value	Lowest possible value
(a)	6 kg		
(b)	17 seconds		
(c)	200 km		
(d)	€12		

24.4 Checking your answers

1 Use + and − together with the three numbers given to make three different sums.
(a) 45 27 18 **(b)** 16 83 67 **(c)** 100 21 79 **(d)** 56 18 74

2 Use × and ÷ together with the three numbers given to make three different sums.
(a) 5 9 45 **(b)** 7 42 6 **(c)** 112 8 14 **(d)** 56 1288 23

3 When estimating the answer to a calculation using whole numbers you can use the following method.
If the number has one or two digits, round it to the nearest 10.
If the number has three digits, round it to the nearest 100.
Etc.
Use this method to find estimates of the following calculations.

(a) $668 + 149$ **(b)** $1075 - 273$ **(c)** $897 + 523 - 166$
(d) $7 + 973 - 606$ **(e)** $18 + 9742 - 190$ **(f)** 265×34
(g) 389×214 **(h)** 5185×31 **(i)** $4051 \div 8$
(j) $15\,672 \div 41$ **(k)** $92\,385 - 27\,441$ **(l)** $6\,458\,023 \div 278$

4 A book costs €6.93.
Estimate, by rounding, how much 22 identical books will cost.
Check your estimate by working out the answer using a calculator.

5 A manufacturer needs to make 10 715 cars to fill an order.
She has 3599 cars already made and in stock.
(a) Estimate, by rounding, how many more cars need to be made.
(b) Use your calculator to find exactly how many cars need to be made.

6 Use estimation to find out which of the following answers are correct and which are incorrect.
(a) $275 \times 992 = 272\,800$
(b) $96\,594 \div 51 = 18\,940$
(c) $\sqrt{96.3} = 31.03$
(d) $\dfrac{118.7 - 15.44}{18.95} = 5.45$

25 Probability

25.1 Probability scales

1 Choose one of these words or phrases to describe each of these events:

Certain   Very likely   Likely   Evens   Unlikely   Very unlikely   Impossible

  (a) You will eat something in the next 24 hours.
  (b) You will watch TV today.
  (c) There will be a general election this year.
  (d) A Third Division club will win the FA Cup next season.

2 Amy picks one of these number cards without looking.

Certain   Very likely   Likely   Evens   Unlikely   Very unlikely   Impossible

Which of the words or phrases above best describes the probability that she picks a:
  (a) 1   **(b)** 2   **(c)** a number less than 10
  (d) a number more than 10   **(e)** a number which is even?
  (f) Make a copy of this probability scale. Mark the events **(a)** to **(e)** on it.

3 These ten socks are in a drawer.

What is the probability that, without looking, you pull out of the drawer:
  (a) a plain grey sock   **(b)** a stripy sock
  (c) a spotty sock   **(d)** a plain grey or a plain white sock?
Give your answers as fractions.

4 This is a pack of 25 Zenner cards.
They are used to test mind reading.
The pack is shuffled and a card taken at random.
What is the probability that the card:
  (a) has a circle on it
  (b) has a circle or square on it
  (c) does *not have* a star on it?
  (d) Jude picks a star on her first go.
    She does not put the card back.
    She shuffles the pack again and picks another
    card at random.
    Which is it more likely to be, a circle or a star?
    Explain why.

25.2 Fair games and estimating probabilities

1 A fair ten-sided die is numbered from 1 to 10.
- **(a)** Andy throws the die 50 times.
  About how many times would he expect to get a 10?
- **(b)** James and Tom are playing a game with the ten-sided die.
  James wins if he gets a score of 3, 5, 7 or 9.
  Tom wins if he gets an even number.
  Is the game fair? Use probability to help you decide.

2 Aysha threw a fair six-sided die 60 times. She drew a bar chart of her results.
- **(a)** Which shape of graph is most likely to show her results?
  Give a reason for your answer.

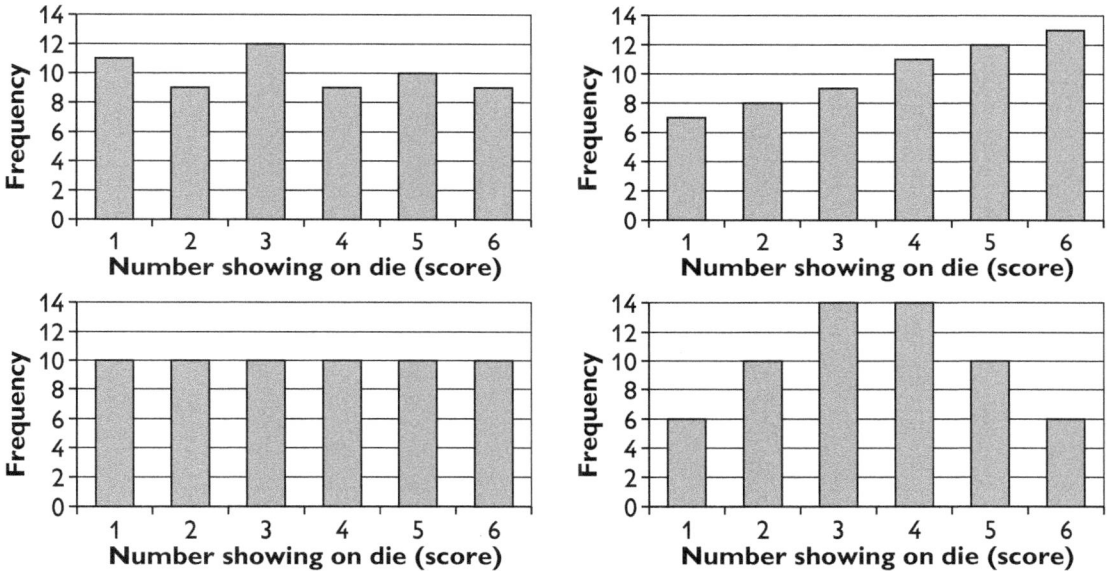

- **(b)** Make a rough sketch of the bar chart you would expect if she threw the die 60 000 times.

3 In a game of chance a person has to pick a cup with a ball underneath.
(Players can't see through the cups!)
- **(a)** Here are two games. Which one gives the player the better chance of winning?
  (i)
  (ii)

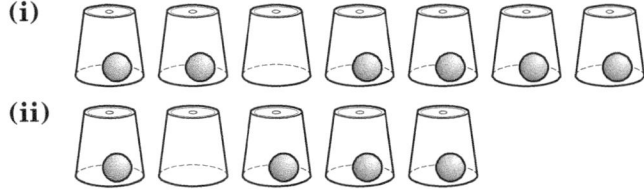

- **(b)** Use probability to explain how you arrived at your answer.

25.3 Combined events

1 This table shows the higher score when two dice are thrown. (Equal scores count as higher.)

(a) Copy and complete the table.
(b) Which is the most likely 'higher score'?
(c) Make a table showing the probability of each of the possible 'higher scores' occurring.
(d) In a game Bill and Ben use the 'higher score'. If the 'higher score' is even Bill wins; if it is odd Ben wins. Is it a fair game? Show how you arrived at your answer.

Second die						
6	6	6	6	6	6	6
5	5	5	5			
4	4	4				
3	3					
2	2					
1						
	1	**2**	**3**	**4**	**5**	**6**

First die

2 Here are two sets of cards, black and grey. One card is chosen at random from each set.

(a) Make a table showing all the possible outcomes.
(b) Use your table to help you find these probabilities:
  (i) both cards showing the same number
  (ii) getting a total of 6
  (iii) getting a total of 4 or more.

| 1 | 2 | 3 |

| 1 | 3 |

3

When two coins are thrown there are just three outcomes. These are 'Heads and Heads' 'Tails and Tails' and 'Heads and Tails'

So the probability of getting 'Heads and Tails' is one third

(a) Get two coins and test to see if what Mark says is reasonable.
Do you agree with him?

(b) Check his working by showing all the possible outcomes in a table.

4 Draw a table to show all the possible outcomes when throwing a four-sided die (labelled 1 to 4) and an eight-sided die (labelled 1 to 8) and adding the scores. Use your diagram to answer these questions.

(a) How many 'doubles' are there?
(b) What is the probability of getting a double?
(c) How many ways are there of getting a total of 6?
(d) What is the most likely total and what is its probability of occurring?
(e) What is the probability of getting a total greater than 5?

26 Transformations

26.1 Rotation

For this exercise you will need some squared paper.

1   Shape A can be rotated through 90°, 180° and 270° to form a pattern.

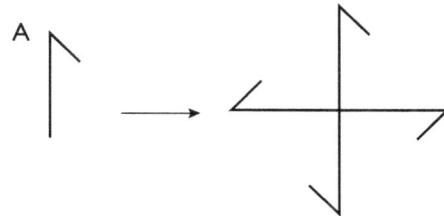

On squared paper, draw a simple shape and rotate it about a point to form your own pattern.

2   Draw axes from −5 to 5 for x and y. Plot the points (1, 1) (1, 4) (2, 4) (2, 3) and (1, 3). Join them to form a flag A.
   (a)   Rotate A through 90° clockwise about (0, 0). Label the image B.
   (b)   Rotate A through 90° anticlockwise about (0, 0). Label the image C.
   (c)   Describe the movement that would map B onto C.

3   Describe fully the rotation which maps:
   (a)   shape A onto shape B
   (b)   shape A onto shape C
   (c)   shape B onto shape C.

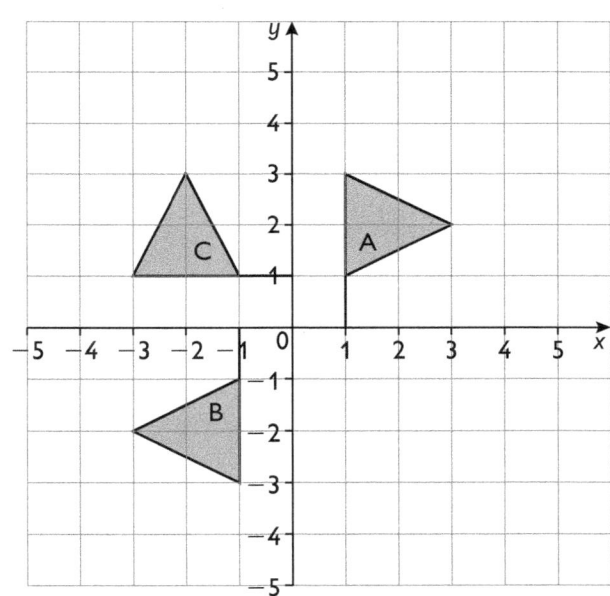

4   Draw axes from −4 to 4 for x and y.
   Draw shape A as in question 3.
   (a)   Rotate shape A through 180° about (1, 0). Label the image B.
   (b)   Rotate shape A through 180° about (0, 1). Label the image C.

5   This star may be made by rotating one point  about the centre of the star.
   What angle of rotation is needed to get to the next point of the star?

26.2 Translation

For this exercise you will need some squared paper.

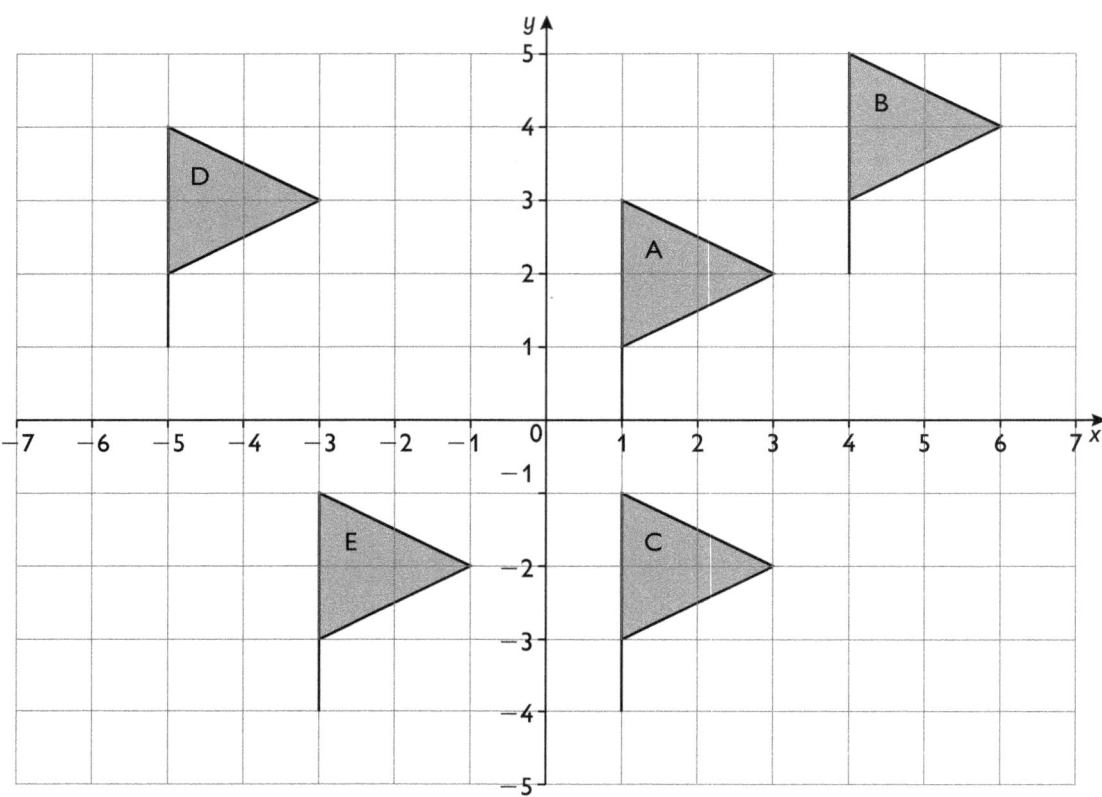

Say how many squares these flags are translated in each direction when:
(a) A moves onto B	**(b)** A moves onto C
(c) A moves onto D	**(d)** A moves onto E
(e) E moves onto C	**(f)** C moves onto E
(g) D moves onto C	**(h)** C moves onto B.

 Draw axes from −6 to 6.
Plot the points (1, 1) (1, 4) (2, 4) (2, 3) and (1, 3). Join them to form a flag A.
(a) Translate A by 3 units to the right and 1 unit up. Label the image B.
(b) Translate A by 5 units to the left and 2 units up. Label the image C.
(c) Translate A by 2 units to the left and 4 units down. Label the image D.
(d) Describe fully the transformation that maps C onto D.

3 Draw axes from −6 to 6 for *x* and *y*.
Plot and join triangle A with vertices at (1, 0) (3, 0) (1, 4).
(a) Translate A by 3 units to the right and 1 unit up. Label the image B.
(b) Translate B by 5 units to the left and 4 units down. Label the image C.
(c) Describe the single transformation that maps A directly onto C.
(d) How can you work out the answer to **(c)** without drawing B and C?

26.3 Reflection

For this exercise you will need some squared paper.

1 Copy this letter L onto a co-ordinate grid.
- **(a)** Reflect L in the x axis. Label the image M.
- **(b)** Reflect L in the line $x = -1$. Label the image P.
- **(c)** Describe the transformation that maps M onto P.
- **(d)** Describe the transformation that maps P onto M.

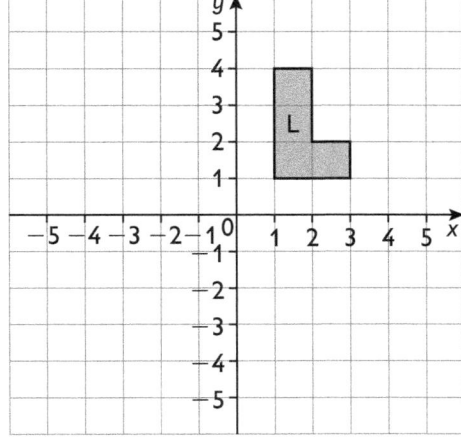

2 Copy Q onto a co-ordinate grid.
- **(a)** Reflect Q in the y axis. Label the image R.
- **(b)** Reflect Q in the line $x = 3$. Label the image S.
- **(c)** Describe the transformation that maps R onto S.

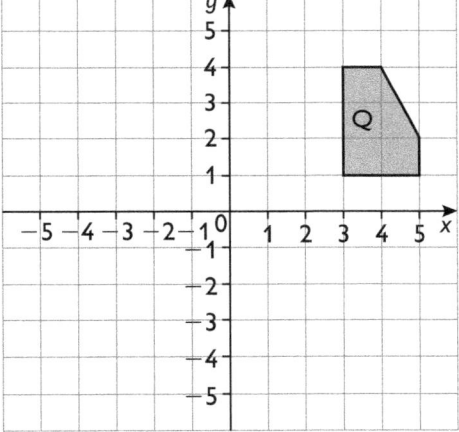

3 Draw axes for x and y from 0 to 6.
Plot and join the points (2, 2) (3, 0) and (6, 1).
Label this triangle A.
- **(a)** Draw the line $y = x$.
Reflect A in the line $y = x$. Label the image B.
- **(b)** Reflect B in the y axis. Label the image C.
- **(c)** Describe the transformation that maps A onto C.

4 **Investigation**

Draw axes from -10 to 10.
(a) Draw a flag A.
Reflect A in the y axis. Call the image B.
Reflect B in the line $x = 4$. Call the image C.
Describe fully the transformation that maps A onto C.
(b) Now reflect A in the x axis and then reflect the image in the line $y = 5$.
What transformation maps A onto the final image?
Here are some ideas for extending your investigation.
- Try reflections in other pairs of parallel lines. State a general rule.
- What happens if you have reflections in perpendicular lines?
- What happens if you have three reflections in parallel lines?

27 Operations in algebra

27.1 Brackets

1 Work out the following.
(a) $(2 + 6) \times 10$ (b) $(5 - 2) \times 7$ (c) $7 \times (4 + 2)$
(d) $(6 - 4) \times 5$ (e) $9 \div (4 - 2)$ (f) $(6 + 9) \div 5$
(g) $(7 - 3) \times (6 + 2)$ (h) $(4 - 6) \times (8 - 3)$ (i) $(5 - 9) \div (1 + 3)$

2 Find the value of the following.
(a) $6 + 7 \times 2$ (b) $4 - 3 \times 8$
(c) $4 - 2 \times 3 + 1$ (d) $7 + 2 \times 9 + 6$
(e) $12 + 8 \div 2$ (f) $10 - 12 \div 3 + 1$

3 Write brackets so that these calculations are correct.
(a) $6 + 7 \times 2 = 26$ (b) $4 - 3 \times 8 = 8$
(c) $4 - 2 \times 3 + 1 = 8$ (d) $7 + 2 \times 9 + 6 = 87$
(e) $7 + 2 \times 9 + 6 = 37$ (f) $7 + 2 \times 9 + 6 = 135$

4 Find the value of the following.
(a) $4(x + y)$ when   (i) $x = 3$ and $y = 4$ (ii) $x = 5$ and $y = 1$
                       (iii) $x = 6$ and $y = -4$ (iv) $x = 7$ and $y = -7$
(b) $5(2x - y)$ when   (i) $x = 3$ and $y = 4$ (ii) $x = 5$ and $y = 1$
                        (iii) $x = 3$ and $y = -4$ (iv) $x = -2$ and $y = -4$
(c) $(x + 3y + 2z) \div 5$ when   (i) $x = 3, y = 4$ and $z = 5$ (ii) $x = 6, y = 1$ and $z = 3$
                                   (iii) $x = 8, y = 4$ and $z = 3$ (iv) $x = 0, y = 1$ and $z = -2$

5 Write the following without brackets.
(a) $3(x + y)$ (b) $4(x - y)$ (c) $5(a + b + c)$
(d) $2(x + 3y)$ (e) $7(a - 2b)$ (f) $6(3p - 2r)$
(g) $8(3m + 2n - p)$ (h) $4(2x - 10y)$ (i) $11(2a + 3b - 5c)$

6 Write the following using brackets.
(a) $5x + 5y$ (b) $7a - 7b$ (c) $3m - 3p$
(d) $4a + 4b - 4c$ (e) $11x - 11y + 11z$ (f) $2a + 4b$
(g) $3x + 6y$ (h) $6x + 2y$ (i) $5a - 20b$

7 Activity

This diagram uses areas of rectangles to show $2(x + y) = 2x + 2y$.

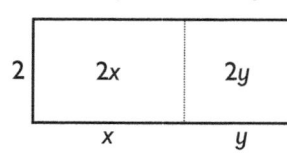

Draw diagrams to show
(a) $4(a + b) = 4a + 4b$ (b) $3(c + 2d) = 3c + 6d$ (c) $2(x - y) = 2x - 2y$.

27.2 BIDMAS

1 Work out the following.
(a) $3 \times 8 + 2$ (b) $3 + 2 \times 9$ (c) $3^2 + 1$
(d) $3^2 \times 4$ (e) 5×2^3 (f) $5 + 2 \times 3^2$
(g) 6×10^2 (h) $7 + 10^2$ (i) $9 + 7 \times 10^2$
(j) $6^2 + 3^2$ (k) $5 \times 10^3 + 4 \times 10^2$ (l) $5^3 + 8 \times 10^2$

2 Work out the following, showing the steps in your working.
(a) $12 - 6 \times 2$ (b) $9^2 + 2 \times 5$ (c) $6 \times 3^2 + 2$
(d) $6 \times (3^2 + 2)$ (e) $100 - 5 \times 2^2$ (f) $(100 - 5) \times 2^2$
(g) $9 \times 10 \div 2$ (h) $6 + 4^2 \div 2$ (i) $(6 + 4^2) \div 2$
(j) $40 \div (3^2 + 1)$ (k) $(1 + 4 \times 2)^2$ (l) $(2^3 - 3)^3$

3 Find the value of the following.
(a) $(x + y)^2$ when   (i) $x = 3$ and $y = 5$ (ii) $x = 6$ and $y = 4$
                   (iii) $x = 5$ and $y = -3$ (iv) $x = -4$ and $y = -1$
(b) $x^2 + y^2$ when   (i) $x = 3$ and $y = 5$ (ii) $x = 6$ and $y = 4$
                   (iii) $x = 5$ and $y = -3$ (iv) $x = -4$ and $y = -1$
(c) $(x + y) \div 2$ when   (i) $x = 3$, and $y = 5$ (ii) $x = 12$ and $y = -4$
                   (iii) $x = -5$ and $y = -1$ (iv) $x = 3$ and $y = -3$

27.3 Collecting like terms

1 Group the like terms together in these lists.
(a) $+2x, +3y, +6x, -2y, +4x$ (b) $+4x, +7, +6y, -3x, +5, -8y$
(c) $+4, -a, +5, +3b, -2a, +6b, -3$ (d) $+5c, -4m, +6s, +c, -3m, -2c$

2 Tidy up the following.
(a) $2a + 5a$ (b) $7b + 2b$
(c) $3c + c + 2c$ (d) $3d + 6d - 2d$
(e) $4e - 7e + 2e$ (f) $3f + 6f - 5f + f$
(g) $3x + 5y + 2x + y$ (h) $5x - 2y + 5y + 9x$
(i) $4x + 7 - 2x + 5$ (j) $6m - 3p - 5m + 2p$
(k) $16r - 9 + 3r - 5$ (l) $5c + 9d - 4e - 3d - 8d + 2c - 7e - d$

3 Mehra is organising a trip to the pantomime with her neighbours.
An adult ticket costs €a and a child's ticket €c.
(a) Write down the cost, in €, of the tickets for each family.
(b) Find the total cost in € for all these families in Mehra's group.
(c) For a group of more than 20 people, $a = 8$ and $c = 3$. Find the total cost using these prices for all Mehra's group.

Pantomime Trip

	Adults	Children
Deans	2	5
Patels	4	2
Jones	1	1
Saids	3	4

4 Tidy up the following. Then use brackets if you can for the final answer.
(a) $3x + 4y - 2y - x$ (b) $7x + 4y - x + 2y$ (c) $15x - 7y - 3x - 5y$
(d) $6x + 7y + 2x - 4y$ (e) $3x + 2y - 5x + 3y$ (f) $6a - 2b + c - 4a + c$
(g) $3(d + e) + d - 7e$ (h) $4(x - 3y) + 9x + 7y$ (i) $5(a + 3b) - 2a + 4b$
(j) $3(a + 2b) + 4(2a - b) - 9a$ (k) $3(a + 2b) + 4(a - b) + 2(3a - 5b) - 5b$

27.4 Inverse operations

1

(a) Jim thinks of the number 4. What is his answer?
(b) Petra's answer is 22. What number did she think of?

2 Here are some LOGO instructions.

    RIGHT 90
    FORWARD 25
    LEFT 90
    BACK 60

Write instructions, from where the turtle is facing now, to get it back to the start and facing in its original direction.

3 Solve these equations. Set out each step fully.
(a) $x + 9 = 12$ (b) $4x = 24$ (c) $x - 6 = 4$
(d) $\dfrac{x}{3} = 5$ (e) $2x = 7$ (f) $2x + 1 = 7$
(g) $3x - 2 = 10$ (h) $4x + 1 = 21$ (i) $5x - 3 = 32$
(j) $7x + 2 = 44$ (k) $6 = 2x - 4$ (l) $9 = 5x + 4$

4 Solve these equations. Set out each step fully.
(a) $2x + 4 = x + 9$ (b) $4x + 1 = x + 7$ (c) $4x - 6 = 3x + 1$
(d) $12x + 1 = 8x + 9$ (e) $7x + 5 = 2x + 15$ (f) $2x + 35 = 7x$
(g) $3x - 12 = x + 6$ (h) $4x + 3 = 2x + 21$ (i) $15x - 3 = 3x + 45$
(j) $7(x + 2) = 42$ (k) $3(x - 4) = 15$ (l) $2(3x - 7) = 4$
(m) $5(4x - 3) = 25$ (n) $4(2x + 7) = 76$ (o) $3x + 8 = 2$
(p) $4x - 1 = 7x + 11$ (q) $3(x + 2) = x + 8$ (r) $5x - 7 = 2(3x - 6)$

5 Check your answer to question 4 by substituting the values you found and seeing whether both sides of the equation give the same result.